T3-AKC-488

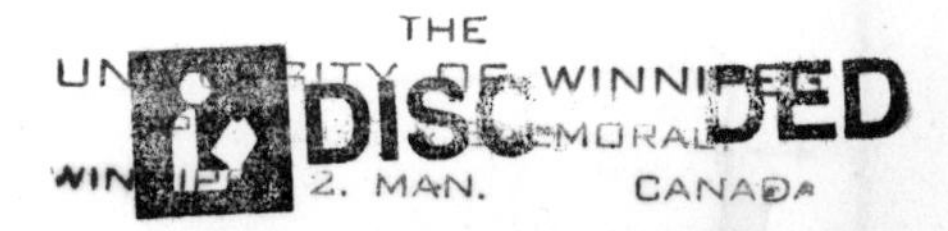

*Settlement & Encounter*

Sir Archibald Grenfell Price

# *Settlement & Encounter*

GEOGRAPHICAL STUDIES PRESENTED TO

## *Sir Grenfell Price*

*Edited by Fay Gale*
*and Graham H. Lawton*

*Melbourne*
OXFORD UNIVERSITY PRESS
LONDON WELLINGTON NEW YORK
1969

*Oxford University Press, Ely House, London, W.1*

GLASGOW NEW YORK TORONTO MELBOURNE WELLINGTON
CAPE TOWN SALISBURY IBADAN NAIROBI LUSAKA ADDIS ABABA
BOMBAY CALCUTTA MADRAS KARACHI LAHORE DACCA
KUALA LUMPUR SINGAPORE HONG KONG TOKYO

*Oxford University Press, 7 Bowen Crescent, Melbourne*



*First published 1969*

SBN 19 550085 7

*Registered in Australia for transmission by post as a book*

PRINTED IN AUSTRALIA BY HALSTEAD PRESS, SYDNEY

# *Foreword*

*by*

THE RT HON. SIR ROBERT MENZIES
KT. C.H. PRS. QC.

Archie Grenfell Price is a geographer and an Australian historian. He is, and has been for a long time, a good friend of mine. His scholarship is of an enviable kind, for it is the adornment of a most civilized mind and a most genial character.

It is a joy to me to know that his students and colleagues are publishing a book of geographical studies in his honour.

In the world of politics, which Archie invaded for too brief a period, reputations are sometimes easily made and even more frequently easily lost. But there are some things which have permanency in them. A scholar's reputation is not assessed by public opinion polls, but by the judgement of his peers.

How great a source of pleasure it is for those of us who know and justly understand Archie to know that a book is being published in his honour; a book which will not smack of the propaganda office but will give him, through the words of the contributors, a just and enduring memorial.

On reflection, I feel I should cancel that word 'memorial'. It sounds a little obituary. And the fact is that Archie will probably outlive me, knowing all the time that I was his friend and admirer.

*Melbourne 1968*

# *Acknowledgements*

We should like to thank two members of the staff of the Department of Geography at the University for their assistance in the preparation of this book. Dr Michael Williams helped with a great deal of the planning and script reading and Mr Max Foale prepared all of the maps and diagrams. Mrs V. M. Benson, of the Barr-Smith Library, prepared the bibliography of Sir Grenfell's publications.

# *Contents*

*Foreword* v

*Sir Grenfell Price*: AN APPRECIATION xiii
ANN MARSHALL, M.SC., M.A.

The Spread of Settlement in South Australia 1
MICHAEL WILLIAMS, B.A., PH.D.

Climate and Man in North-western Queensland 51
F. H. BAUER, B.S., M.A., PH.D.

A Changing Aboriginal Population 65
FAY GALE, B.A., PH.D.

Apartheid: Background, Problems and Prospects 89
R. K. HEFFORD, M.EC., M.A.I.A.S.

Problems of Vegetation Change in Western Viti Levu, Fiji 115
G. ROSS COCHRANE, M.A., B.SC.

Jet Age Medical Geography 149
BRIAN MAEGRAITH, C.M.G., M.A., M.B., D.PHIL., D.SC., F.R.C.P., F.R.C.P.E.

Some Non-nutritive Functions of Food in New Guinea 173
DAVID LEA, B.A., PH.D.

Australia in New Guinea: None So Blind . . . 185
DIANA HOWLETT, B.A., PH.D.

A Macrogeography of Western Imperialism: Some Morphologies of Moving Frontiers of Political Control 213
D. W. MEINIG, B.S., M.A., PH.D.

*A Bibliography of the work of Sir Grenfell Price* 241

*Notes on Contributors* 245

*Index* 247

# Illustrations

## PLATES


| | | |
|---|---|---|
| *Frontispiece* | Sir Grenfell Price | *between pages* 124-6 |
| PLATE 1 | Hill country changed by fire, Yavuna, Fiji | |
| PLATE 2 | Bush, Nausori Highlands, Fiji | |
| PLATE 3 | Small plot of Fijian agriculture | |
| PLATE 4 | Vegetation in Tavua basin, Fiji | |
| PLATE 5 | Slips on burnt hills, Yavuna, Fiji | |
| PLATE 6 | Ba Closed Area, a fire-induced desert, Fiji | |
| PLATE 7 | Farming in Ba Basin, Fiji | |
| PLATE 8 | *Qato* in the eroded Ba Closed Area, Fiji | |


## FIGURES


| | | |
|---|---|---|
| Fig. 1 | Generalized terrain regions in South Australia | 5 |
| Fig. 2 | Average annual rainfall in South Australia | 6 |
| Fig. 3 | The soil landscapes of South Australia | 7 |
| Fig. 4 | The vegetation of South Australia | 8 |
| Fig. 5 | Spread of settlement in South Australia, 1836-49 | 12 |
| Fig. 6 | Spread of settlement in South Australia, 1850-59 | 16 |
| Fig. 7 | Spread of settlement in South Australia, 1860-68 | 20 |
| Fig. 8 | Five aspects of South Australian land settlement | 23 |
| Fig. 9 | Spread of settlement in South Australia, 1869-79 | 27 |
| Fig. 10 | Spread of settlement in South Australia, 1880-92 | 34 |
| Fig. 11 | Spread of settlement in South Australia, 1893-1919 | 39 |
| Fig. 12 | Spread of settlement in South Australia, 1920 onwards | 44 |
| Fig. 13 | Settlement in North-western Queensland | 53 |
| Fig. 14 | A comparison of Aboriginal and white population change, Australia, 1870-1964 | 71 |
| Fig. 15 | Changing age structure of Aborigines between 1930 and 1964, compared with the age structure of other Australians, 1961 | 78 |

Fig. 16 Age-specific fertility rates of Aboriginal females, 1945-59, compared with those of other Australian females in 1954 83
Fig. 17 Location map, Viti Levu, Fiji 116
Fig. 18 Vegetation map, Viti Levu, Fiji 118
Fig. 19 Location map of Nausori Highlands, Yavuna, and Koroba Range, Western Viti Levu, Fiji 125
Fig. 20 Cross-section from Koroba through Yavuna area to Nausori Highlands, Fiji 127
Fig. 21 Representative slopes and frequency of slips, Yavuna, Fiji 131
Fig. 22 Valley profiles in the Yavuna area, Fiji 133
Fig. 23 Sparse root 'network' of grass communities, Fiji 135
Fig. 24 World distribution of malaria and smallpox 158
Fig. 25 Thrusts from European Christendom on the eve of the fifteenth century 215
Fig. 26 Hypothetical patterns of the continental empire of Spain 217
Fig. 27 Hypothetical sea empire 223
Fig. 28 Segmentation by sea empire rivalries, Guinea Coast, *c.* 1700 223
Fig. 29 Hypothetical riverine empire 226
Fig. 30 Hypothetical colony of a settler empire 230
Fig. 31 Hypothetical patterns of a nationalistic empire 235
Fig. 32 Hypothetical patterns of culture areas and imperial areas 237

# *Sir Grenfell Price*

## AN APPRECIATION

THE MEASURE of a man's value in his chosen field must be made, not only in terms of his own personal contribution, but also in terms of the stimulus he gives to others. On both these counts, Grenfell Price has a high place in Australian geography. His own interests have been wide and varied, and his work has been significant in many fields. In addition, his infectious enthusiasm, his light-hearted disregard for conventional ideas and his warm and friendly personality have made him a lecturer to be remembered. Many of his students who have gone on to distinction in their own fields will say they are geographers because of Grenfell Price.

His early interest in exploration and discovery may well reflect his childhood in a period when a great deal had still to be learned about South Australia and the dividing line between surveying and exploration was very nebulous. His uncle, Captain W. N. Goalen, R.N., surveyed much of the South Australian coast, and against the backdrop of his experience moved many interesting explorer-surveyors such as George Goyder. By the time he went to Oxford to do a History Honours degree, he had a particular interest in the section of the work which dealt with exploration and mapping, and in the training for his Diploma of Education he made his first contact with geography as a separate discipline. On his return to Australia as a master at St Peter's College he was given the responsibility of teaching geography to senior forms. It is typical of Dr Price that he regarded this as a challenge, and it led him to a life-long interest in the subject.

Geography in the 1920s, especially in Australia, was greatly dominated by studies of the physical landscape, with a very simple and uncritical determinism as something of an afterthought. Dr Price's historical background was a balance to this strongly physiographic viewpoint. Although he accepted the current ideas, he was always more interested in individuals than in 'controls', and his writing always warmed when he discussed the influence of man rather than the dominance of 'nature'. His important works on the historical geography of South Australia date from the years when he became the Master of the first residential University college in

Adelaide, St Mark's. In his own words, he 'wrote such books as one could spare the time for on the historical geography of South Australia'.

In one of the earliest of these books (1924), entitled *The Foundation and Settlement of South Australia, 1829-1845*, Dr Price maintained that the early history of the State had been misinterpreted. Not enough attention in his view had been given to the geographic and economic aspects of the story. Many colonists, for example, misunderstood the peculiarities of the Mediterranean environment and agricultural production therefore did not keep pace with the influx of population. Much of the limited capital of the colony had to be expended on imports of necessary food supplies.

This work on the history of the settlement of South Australia was soon followed by a series of portraits of its founders. From collections of official and private papers in the South Australian Archives and from Mr C. H. Angas, Dr Price obtained valuable source material which became the basis for his book *Founders and Pioneers of South Australia*, which he hoped would stimulate 'interest in the early leaders of the province' and their endeavours to establish a colony under insuperable difficulties. The importance of the contribution of the explorer and the adventurous settler in the early history of the State is brought out clearly, and a fine tribute is paid to Wakefield who was not only the father and chief founder of South Australia but an Imperial statesman to whom Australia as a whole owes much of its free settlement.

These books were regarded as the authoritative work on the period, and formed a major part of the work submitted to the University of Adelaide for the degree of Doctor of Letters, which he was awarded in 1932.

Work on the historical documents of Northern Territory administration led to a growing interest in the settlement of tropical areas by white people, and in 1929 he made his first trip to tropical areas, Java, Burma and Ceylon. Dr Rafael Cilento was at this time developing his unconventional views of white settlement in the tropics, and contacts with him stimulated Dr Price's interest still further. In 1932 he was awarded a Rockefeller Service Fellowship to study the history of white settlement in the tropics. He journeyed through the arid centre of Australia to Darwin to get some first hand impressions of the country, and his assessment of the potential for development of the Australian winter-dry tropics was highly realistic, and far from the incurable optimism of the time. In fact his statement that there would be little settlement until 'the more attractive parts of Australia are sufficiently closely settled to create population pressure' sounds like the agricultural economists of thirty years later. This journey confirmed a growing interest in the problems of white and coloured races in the same area.

At this time, every failure in tropical settlement by whites was attributed directly to the climate, and it was Dr Price's refusal to accept the dominant point of view, simply because it was the dominant point of view, which led him to his valuable reassessment of the tropical environment. With a background of history and a keen interest in the different historical backgrounds of the different tropical islands, Dr Price started his project unimpressed by any pre-conceived ideas about the unsuitability of white races for tropical living. He wanted to find out 'why the Spanish had succeeded in certain areas . . . and why in almost every area the North European whites had failed'. In 1933 he and his wife visited Cuba, Jamaica, Panama, Costa Rica, Curaçao, Bonaire, St Thomas, Saba and St Kitts—and considered each one as a potentially different problem. More and more, the importance of disease and of a competitively low standard of living among the coloured races emerged as the principal factors in the decline of the white population. The observations of these months were summarized in *White Settlers in the Tropics* which the American Geographical Society published in 1939. In many ways, this book ranks as one of Dr Price's most important contributions to geography. It has influenced thinking about tropical problems, and opened up many new lines of interest in acclimatization. While he was working in the remote island of St Kitts, he received notification that he had been awarded the C.M.G. for services as chairman of the Emergency Committee of South Australia during the Australian financial depression.

On his return from the Caribbean, Dr Price had a chance to experience the conditions of Australian exploration at first hand, when he was given charge of an expedition to search for traces of the explorer Leichhardt, lost in 1848. The party left Adelaide in August 1938 and went down the dry bed of the Finke River from Mt Dare station to check the authenticity of 'remains' which had been reported on the border of the Simpson Desert. Extensive digging showed traces of a very old camp, and included a Maundy threepence which was dated the year Leichhardt left England, but the results had to be described as 'inconclusive'. However, his close association with the scientific work carried out by Dr Campbell and Mr Mountford stimulated Dr Price to further interest in Aboriginal occupation.

In 1941 Dr Price became the Liberal Member for Boothby in the Federal Parliament, and in his own words 'kept Menzies in office for some months'. In his term at Canberra, he worked with Arthur Caldwell, and shared a room for some time with Harold Holt. He also formed the close association with Sir Robert Menzies which is reflected in the warmth of Sir Robert's Foreword to this book.

The years of the Second World War saw a growth of interest in geography as an academic subject at the University of Adelaide, and a visit from the late Griffith Taylor led to the appointment of Dr Price in 1949 to develop a third year course in the subject. He held this position until 1957, and in those years made his dynamic contribution to University geography by the medium of his lectures and his personal contacts. His range of interest was spectacular, and his verve and enthusiasm were unlimited. (Asked what it was like to work with Dr Price in those years, I replied with feeling that it was like being the tail of a comet.) I think there is no doubt that he was responsible in no small way for the development of the then part-time department into an Honours School in 1951.

Dr Price left the Geography Department at the end of 1957, when the rapid growth of classes called for full-time staff. His life was already crammed with other activities. He had been a member of the Council of the University of Adelaide since 1925—a position he held until 1962. The Council Minutes record a recognition of his professional activities and add 'The Council and the University as a whole are deeply indebted to Dr Grenfell Price for his vigorous service maintained over so many years.' He was a member of the Faculty of Arts from 1926 to 1965, and its Dean in 1951-52. He is a Fellow of the Royal Geographical Society of London, and he was appointed to the Council of the Royal Geographical Society of Australasia (S.A. Branch) in 1925, served as its President in 1937-38 and was awarded its John Lewis Gold Medal for his geographic research in 1949. He has been a member of the Libraries Board of South Australia since its inception in 1940 and the Libraries Board wishes me to record 'its deep appreciation of the refreshing outlook, breadth of knowledge, and kindly wisdom which he has brought to its deliberations'.

Behind all these activities was what he would probably regard as his main occupation—his position as Master of St Mark's

College. In the preface to *A History of St. Mark's College* (which Dr Price himself wrote in 1967) the Bishop of Adelaide says 'there is only one thing Sir Grenfell Price's history fails to do—record the unique services which he and Lady Price have rendered to the college.' This preface concludes with a reference to 'the immense debt the College will always owe to Sir Grenfell and Lady Price, who watched over its birth, childhood and coming of age, and who set its feet firmly upon the stage of Australian academic life.' I am quite sure that Dr Price values this tribute as much as any of the honours of his professional career.

With indefatigable interest Dr Price has continued a spectrum of activity which would do credit to a man twenty years younger. He was appointed the secretary of the Humanities Research Council, and in this capacity edited the annual reports and was responsible for the early monographs. He became chairman of the Advisory Board of the Commonwealth Literary Fund. He turned his attention to the editing of important books on geographic subjects—notably *The Explorations of Captain James Cook in the Pacific as told by selections of his own Journals, 1768-1779* and *The Winning of Australian Antarctica—Mawson's 'Banzare' Voyages.* In addition, he wrote the third of his books on white settlement—*The Western Invasions of the Pacific and its Continents*—in which he summarized the stimulating lines of interest which had given so much colour to his lectures. It continues the theme of contrast between 'settler' and 'sojourner' types of colonization, and shows the emergence of Dr Price's interest in the moving frontiers of plants, animals and disease, a theme particularly interesting and important in the Pacific area. To quote one reviewer 'it will make us view our specialized knowledge in a wider context: it is intended to provoke thought and research.' Such has always been the intention of Dr Price's writing.

Although his continued interest in geography is evidenced by several addresses and short papers, and by the publication of his most recent geographic book *The Challenge of New Guinea*, his main energies have been increasingly devoted to the extremely important position of chairman of the National Library. The recent opening of the magnificent building and the tributes paid to the organization which has already amassed over a million books must be regarded also as a personal tribute to his success in this position.

Lady Price shared her husband's interests and accompanied him on almost all his varied travels—she has said she sailed every

Pacific wave with Cook. She added her delicate and ironic humour to his unfailing good temper, and students and staff have always found them approachable to the highest degree and the best possible company.

This is an appreciation of Dr Price's personal influence rather than a detailed review of his work, and there is no place here for discussion of his numerous publications, which are listed in the Bibliography. He says himself 'although my own work looks so haphazard, it all evolved logically.' This logical evolution can be seen in the Bibliography, and in the essays in this book which are arranged to follow the same pattern: Australian school geography; a fusion of historical and geographical themes leading to South Australian historical geography—Williams; historical geography of Northern Australia and white settlement in the tropics—Bauer; the inter-relationships of white and native peoples—Gale and Hefford; the spread of exotic plants and animals—Cochrane; the spread of exotic diseases—Maegraith; the impact of Australian interests in New Guinea—Lea and Howlett; and finally the broad problems of moving frontiers of politics—Meinig. Many of the writers owe their interest in these subjects to their contact with Dr Price's own writing or teaching. As such, the book represents that larger projection of his own interests which is the best evidence of a stimulating teacher and an engaging personality.

ANN MARSHALL

*For*

*A. Grenfell Price*

*Friend and mentor, whose contagious enthusiasm*
*stimulated many students and colleagues*
*to envisage wider horizons*

MICHAEL WILLIAMS

# *The Spread of Settlement in South Australia*

*'The pioneers of South Australia were human in their faults, but on the whole they worked in right directions, and laid the foundations firmly on rock. In spite of initial mistakes, they conquered the great difficulties of a new and unknown environment. . . .'*

A. GRENFELL PRICE *The Foundation and Settlement of South Australia, 1829-1845*

## INTRODUCTION

ONE OF THE most fascinating aspects of the European agricultural colonizations of the nineteenth and early twentieth centuries is the rapidity with which large parts of the earth's surface came to have a new geography. The new organization of land and life often started from an uninhabited, or usually at best, sparsely inhabited landscape, largely, if not wholly unmodified by man. What changes subsequently took place were, therefore, revolutionary in their effect, and because of the level of technology of the time, rapid in their achievement. When compared with the history of human settlement up till then, the outcome was one of stark contrasts within a relatively short space of time as, for example, tree-felling and scrub-clearing obliterated one vegetative cover and was replaced by another; as hydrological features of the landscape were changed by draining and irrigation, and soils were inevitably altered by these and other human activities. Concomitant with these changes of the biosphere came the creation of new settlements and farms, new roads and railways, as well as the introduction of new fauna and flora.

Although each of the stages or periods in the colonization of any such area of the world must ultimately and necessarily be studied in detail so that the fullest understanding can be gained of the changes that occurred, there is a virtue in attempting to present an overall view that traces these far-reaching and rapid changes over large areas with some precision. This is particularly true in Australia, where few such studies exist of the changing geography of any one large area through all its phases of settlement (Bauer 1959, Buxton 1967, Heathcote 1965, Kenyon 1914, 1915).

The historical geographer in this continent constantly finds himself asking the most elementary of geographical questions, such as, where and when colonization took place, how this colonization was undertaken and what effect it had on the landscape.

In South Australia the picture is perhaps brighter than in other States. Excellent work has been done by geographers and historians on the earliest phases of settlement, i.e. from 1836 to 1855 (Price 1924, Pike 1957) and there exists an exemplary study of the colonization of a large part of the State for what may be termed the 'middle years' of settlement, from 1869 to 1884 (Meinig 1962). Nevertheless, the need for a geographical overview is not diminished, for such a framework will provide a context in which detailed local studies can be placed, and the total picture will be illustrative of themes and phases that could be pursued in the evaluation of the changing geography of other parts of the continent.

This paper analyses the colonization of South Australia by a series of seven maps, employing two criteria to delimit the spread of settlement; the proclamation of cadastral units known as hundreds and the survey of township sites.[1]

The reasons for the choice of these two criteria have been examined elsewhere (Williams, M. 1966a), but suffice it to say here, that the hundreds had to be proclaimed and the rural sections surveyed before any permanent alienation of the land and hence settlement could take place. Therefore, the pattern of hundreds tends to give an exaggerated picture of the absolute amount of land settled at any one time, but a good indication of the relative

[1] A hundred is a roughly rectangular survey unit of approximately 100 square miles, though the great majority ranged between 90 and 130 square miles and extreme sizes were 33 and 304 square miles. They had to be surveyed before any permanent sale and settlement could take place and were a way whereby the State government controlled the pace and direction of settlement. Later, groups of hundreds formed local government and administrative units. The dates of proclamation are taken from records in the Lands Department, Adelaide. The word 'township' is widely used in South Australia for a nucleated settlement having retail and residential functions, though many townships were little more than 'villages' or 'hamlets' and some that were surveyed do not exist today. The dates of survey of government townships are taken from records in the Lands Department, Adelaide, whereas the dates of creation of private townships are based on a variety of sources ranging from local histories and land agents' subdivisional plans to petitions for police protection, a good indication of population nucleation. These, and many other classes of sources, were consulted in the South Australian Archives, Adelaide.

sequence of settlement. The distribution of townships adds precision to the picture by suggesting the presence of a core of well settled land. Railways have been added because they indicate the main patterns of circulation that evolved in the newly colonized areas, and in the later phases of settlement, they were also the spearheads of new expansion.

This study does not seek to reconstruct total geographies of the whole of South Australia for past periods. Rather, the focus of attention is constantly on the newly emerging areas of settlement and little account is taken of the older frontiers of colonization. This intentional emphasis is pursued in order to illustrate the sequence and the rapidity of the colonization process. Also, throughout this study the processes of landscape change are emphasized. Yet, because the interval between each map and phase is small, and because each map incorporates within it the residue of the former phase, it is hoped that the total sequence so presented is not one of isolated and static pictures, but one that does provide an interpretation and explanation of the distributions (Darby 1962), and a continuous view of the marked and rapid changes that occurred from place to place, from time to time in South Australia.

## HISTORICAL SETTING

Any attempt to interpret the distributions shown on Figures 5 to 12 must first take into account the conditions that led to the foundation of South Australia and the principles of colonization which influenced the course of settlement (Price 1924, Pike 1957).

The early nineteenth century was a period of internal economic and social discontent in Britain as the Industrial Revolution got under way; whilst overseas, colonies had been lost (U.S.A.) and other colonial enterprises had failed or were an indifferent success (Cape Colony, Swan River, New South Wales), while others still were bordering on revolt (the Canadian uprisings which finally erupted in 1837). It seemed to some economic and political theorists that both the internal and external problems could be resolved and rationalized by a scheme of orderly colonization which absorbed the poor and unemployed at home, and established a new and stable colonial society overseas.

Of the many people involved in the planning of South Australia, E. G. Wakefield must be singled out as the one person who did more than any other to weld the many theories together into a plan of 'systematic colonization'. The main principles of colonization were firstly, that land should be sold at a fixed minimum price or more, and that the revenue obtained should be used to assist the emigration of new settlers; secondly, that the volume and pace of colonization should be regulated by the amount of land available, and that settlement should expand in contiguous blocks. Implicit in these proposals was the idea of the survey of the land prior to its alienation and an emphasis on family colonization. These principles were aimed primarily against the transportation of convicts and bonded servants and against the problems of squatter occupation, both of which had bedevilled the early years of settlement in all other parts of Australia. The discovery of the southern coast of Australia and the exploration of the River Murray down to its mouth provided the theorists with a locale for their experiment, and in 1836 the first colonists landed on the coast, near what is now Adelaide.

In this summary statement of the origin and early years of South Australia, emphasis has been placed on the early theoretical planning of settlement from afar to produce an ordered and stable society. Even when the colony became self-governing in 1857, the government did not disregard these ideals but remained true to its heritage and took a firm hand in controlling settlement, which was such a conspicuous contrast to contemporary colonizations in other parts of Australia, and of course, elsewhere in the world.

## THE PHYSICAL AND BIOTIC SETTING

The conditions of colonization explain much about the settlement of South Australia; but the land itself, its physique, rainfall, soils, original vegetation and water supply are also important. Despite general governmental control of colonization much was still up to the practically-minded immigrant farmer who saw the country and assessed its potentialities as a place for pioneering. He found a land which, although locally varied, was broadly simple in its features and in his eyes some localities were initially favourable for settlement, whereas other localities were not.

### *The Favourable Areas*

Particularly attractive was the upraised block of land that extends northwards through the centre of South Australia like a spine. This was where first settlement had taken place (Figure 1).

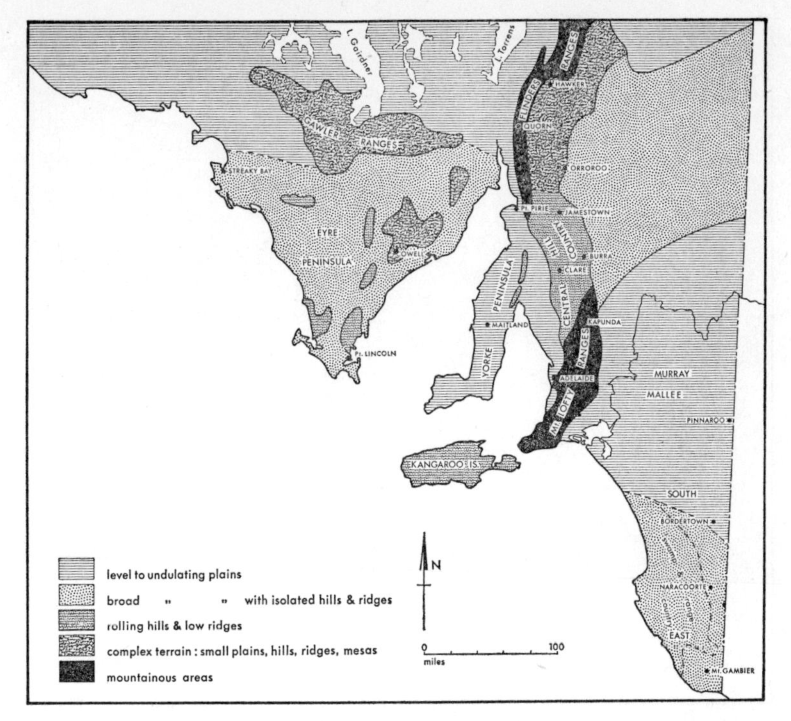


FIG. 1 Generalized terrain regions in South Australia
Based on C. Fenner (1931), A. K. Lobeck *et al.* (1951) and Meinig (1962)

From Fleurieu Peninsula the Mount Lofty Ranges rise to over 1,500 feet, but many broad upper-valley reaches and some fault-formed coastal basins and plains provided suitable locations for settlement. Beyond the Barossa Valley is the Central Hill country where the general highland relief diminishes, and is replaced by a series of longitudinal ridges which separate wide alluvial plains and rolling hills. North of this zone and extending into the arid interior is the high ground and truly mountainous relief of the Flinders Ranges which have an abrupt western scarp, but a broken eastern margin of isolated hills, and of basins and plains.

Relief has a marked effect on the rainfall (Figure 2) which is over 15 inches per annum throughout the greater part of this upland zone, and over 25 inches in the southern core of the Mount

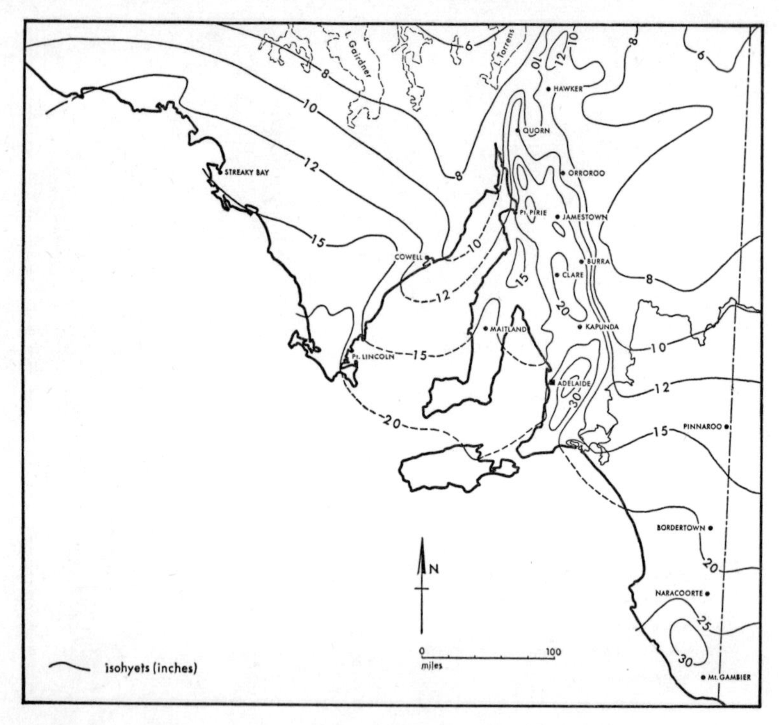


FIG. 2 Average annual rainfall in South Australia
Based on Commonwealth Bureau of Meteorology (1950)

Lofty Ranges. The 'Mediterranean' winter maximum of the rainfall, separated by dry and sometimes searingly hot summers, was an aspect of the environment which took the British immigrant farmers a few years to become accustomed to, and reinforced their preference for the high rainfall areas of the upland block.

The soils, broadly speaking, are fairly fertile, crusty and hard setting loams with a predominantly red or brown subsoil which in the higher rainfall of the southern Mount Lofty Ranges tend to greater podzolization, less fertility, and yellow colouring (Figure 3). Their vegetative cover is savannah woodland of blue gum (*Eucalyptus leucoxylon*) and peppermint gum (*Eucalyptus odorata*), the latter degenerating to a mallee habit as the rainfall declines northwards beyond the Central Hill country, and changing to open grassland on the drier eastern side of the hills near Burra

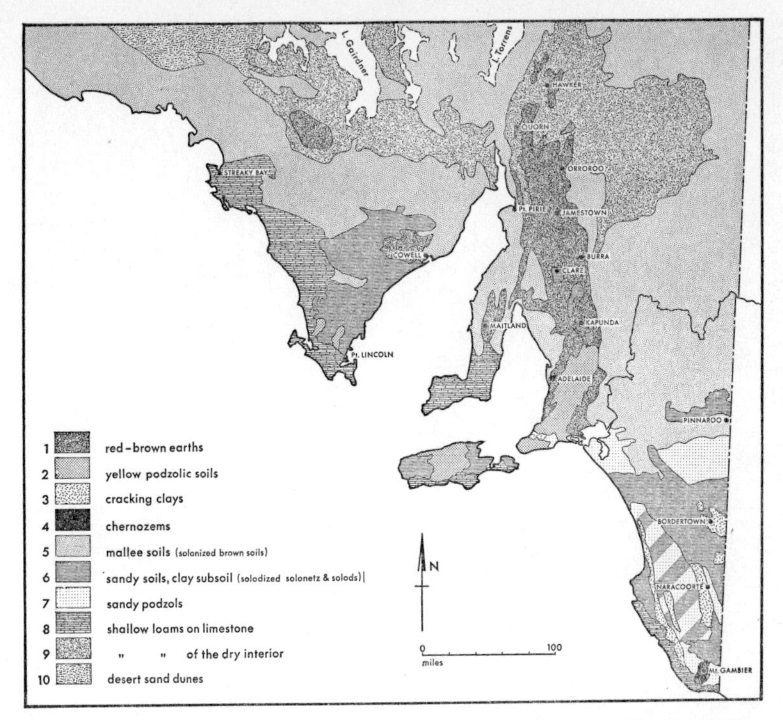


FIG. 3 The soil landscapes of South Australia. Based on K. H. Northcote (1960). The advice of Mr Northcote in the construction of this map is gratefully acknowledged. The groupings on the above map consist of the following soils, the characteristics of which are indicated by the key symbols used in the *Atlas of Australian Soils*. They are:

1 = Dr2·23; Dr2·33. 2 = Dy3·61 and others. 3 = Ug5·11; Ug5·2. 4 = Um6·11; Um6·12. 5 = Gc1·11; Gc1·12; Gc1·22. 6 = Dy5·43; Dy3·42 and others. 7 = Uc2·2; Uc2·3. 8 = Uc6·13; Um6·24. 9 = Um5. 10 = Uc1·2.

(Figure 4).[2] In the higher rainfall zone (over 25 inches per annum) of the Mount Lofty Ranges, the savannah woodland is replaced by a dense and almost impenetrable stringybark forest of *Eucalyptus obliqua*, *baxteri* and *cosmophylla*, which was largely

[2] It should be noted that the boundaries of vegetation regions vary in Williams, R. T. (1955) and Wood (1937), particularly in Eyre Peninsula. However, Figure 4 is based largely on Williams, except where Wood's more detailed mapping produces variations that seem significant to this study. Contemporary descriptions of vegetation, based on J. Backhouse (1843) and J. W. Bull (1884) are collected in Cleland (1928).

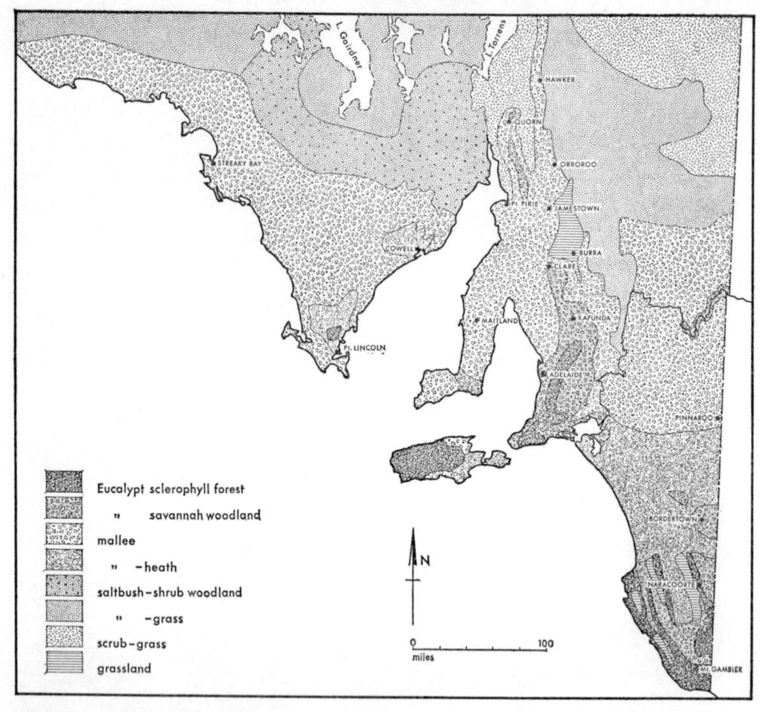


FIG. 4 The vegetation of South Australia
Based on R. T. Williams (1955) and J. G. Wood (1937)

avoided. A fairly similar, though smaller zone of eucalypt forest and woodland on yellow podzolic soils, and some sandier soils, exists on the higher rainfall tip of Eyre Peninsula. The savannah woodland, with its herbaceous and grassy ground cover, provided good grazing for sheep without clearing, and was relatively easy to clear when cultivation was attempted. It is in these woodlands that some of the early episodes of colonization were enacted.

Equally favourable, but in many ways different in character, were parts of the Lower South-East district. Here, a complex topography of low north-west to south-east trending sandy and calcareous ranges separate wide level plains or 'flats', each flat being at a higher level than the one adjoining to the west, the easternmost flat near Naracoorte levelling out to a gently undulating plain of about 200 feet above sea level. Until the present, these flats have been subject to prolonged and extensive flooding because of high rainfall, natural drainage impeded by the low ranges, and excessive ground water, which gravitates from Victoria and rises to the surface in the winter months. But at the southern and eastern margins of this uninviting zone of flooded flats and infertile ridges are the black loamy, almost chernozem, soils developed on the volcanic deposits near Mt Gambier, and the sandier, though moderately fertile soils of the border upland, both areas carrying a savannah woodland vegetation, which merges into open grasslands on the flats in the centre of the region. Both of these areas were of outstanding attraction to early settlers.

Another group of soils favourable for early settlement were the black to dark grey clays of the Bordertown district, which were subject to excessive cracking in the summer and swelling in winter and which produced the typical 'crab-hole' country of the wheatlands of the Wimmera in Victoria, of which this district forms but a small western extension. The potential fertility of these soils when ploughed was not appreciated until the late 1860s. Some of the larger grass-covered flats in the Lower South-East, such as the Mosquito Plains west of Naracoorte, possess similar soils, but periodic flooding makes them far less suitable for cultivation.

*The Less Favourable Areas*

In contrast, there were the areas less favourable for settlement that occupied the remainder of South Australia. Some of these areas, such as the dense eucalypt forests of the Mount Lofty Ranges and the swamps of the Lower South-East, have been mentioned already, but the remaining areas were of a different character and consisted mainly of the flat to gently undulating plains that stretched through Yorke Peninsula, the Murray Mallee, the Upper South-East and the greater part of Eyre Peninsula. The soils of these plains consist of a great variety of sands and loams, the most widespread being the highly calcareous loamy earths (Solonized Brown Soils) which stretch across Eyre Peninsula, occupy most of the north of Yorke Peninsula and the eastern fringes of St Vincent and Spencer Gulfs, and nearly the whole of the Murray Mallee. These 'mallee' soils, as they are sometimes called because of their characteristic vegetation, have many north-west to south-east trending dunes which give local relief, but once their vegetation is removed the soil tends to drift easily. Immediately below the loamy surface, there often occur nodules of limestone, which, along the west coast of Eyre Peninsula and the foot of Yorke Peninsula constitute a solid sheet of limestone, giving terra rosa soils, which hinder cultivation. All these sandy and shallow soils are not naturally fertile; they are deficient in phosphorus and low in organic matter. There is a lack of surface water.

But an even greater obstacle to the initial utilization of these soils was the dense cover of mallee woodland or scrub which formed an almost impenetrable barrier. There was no way of clearing this scrub vegetation except by hand, until the process of rolling down its slender stems was discovered in 1868. This discovery aided settlement and with the conclusive proof, after about 1890, that superphosphate fertilizer could materially alter the basic fertility of the soils, great inroads of settlement were made in the mallee lands.

The seemingly endless expanse of mallee soils and vegetation were relieved in a few places with patches of better soils, for example, the red loams near Cowell in the Eyre Peninsula, and around Maitland in Yorke Peninsula; all of which proved more attractive than the surrounding land to the agricultural pioneers during the period of rapid settlement from 1869 to 1880.

Even more difficult and repelling than the mallee lands were the sandy soils (solodized—solonetz and solods) of the eastern parts of Eyre Peninsula, and the highly podzolized soils of Kangaroo Island and the Fleurieu Peninsula. High rainfall and a prolific vegetation cover of eucalypt woodland in Kangaroo Island and the two peninsulas masked their basic infertility, due to trace-element deficiencies of copper, zinc and molybdenum. Fleurieu Peninsula, in particular, was the scene of early abortive attempts at settlement. The mallee-heath vegetation associated with the lower rainfall of the Upper South-East was quite clearly repelling—the local name of the region, the Ninety-Mile Desert, was indicative of what the pioneers thought of the region. Not until after the Second World War was the trace-element problem of these areas solved. Agricultural settlement then took place. Northwards, and beyond all these difficult areas, lay the true semi-desert with an annual rainfall of below 10 inches, with amorphous sandy and sometimes loamy soils, and with a varied vegetational cover of grasses, saltbush, scrub and stunted mallee, only the better portions of which were used for purely pastoral purposes. These areas merge eventually into the true desert sand-dunes.

From this survey it is apparent that an important feature of early settlement was the isolation of the initially attractive core areas by dry mallee-covered plains, in the case of the Mount

Lofty Ranges and Eyre Peninsula, and by the trace-element deficient Ninety-Mile Desert and flooded swamps in the case of the Lower South-East district. Consequently, once these core areas had become fairly well settled the feeling of isolation was accompanied by one of confinement, and much of South Australian settlement geography is concerned with the breaking down of these barriers of difficult lands as new techniques and inventions either came to hand or were developed specifically to meet the needs of the pioneers on the land.

## BEGINNINGS, 1836-1849

The great and pressing need of the colonists after the creation and survey of Adelaide in 1837 was to get on to the land in order to make themselves self-supporting in food. But because of delays in the survey of the rural sections and because of some initial mismanagement the colonists had to remain in Adelaide, and it was not until 1840 that they were able to start farming in earnest, and were said to be 'going into the interior in shoals' (B.P.P.: No. 394, 1841). The direction of movement from the Adelaide nucleus was twofold; firstly, southwards along the narrow coastal plains and basins, and secondly, across to the eastern side of the Mount Lofty Ranges and then southwards towards the Murray mouth (Figure 5).

The Adelaide plain and the Noarlunga and Willunga Basins further south were covered with alluvial and colluvial soils and were well watered with streams running off the Mount Lofty Ranges. The savannah woodland required little clearing for cultivation and the depasturing of stock was easy. By the end of 1840, 13,513 acres of land were enclosed, of which 1,977 acres were under cultivation, almost wholly wheat. Cultivation at this time did not extend northwards in the coastal plains beyond Port Adelaide because of the presence of coastal swamps and because of the decreasing rainfall (B.P.P.: No. 505, 1843).

It would be impossible to show on a map of the scale of Figure 5 the great number of small agricultural villages and hamlets that grew up in the plains surrounding Adelaide, but by 1840 alone there were 26 'villages' in existence containing 'at least a few houses, the residences probably of a carpenter, a blacksmith or shoemaker' (Bennett 1842: 132), although this description omitted the indispensable hotel, and, later, the general store, which were important in drawing trade from the surrounding country. A

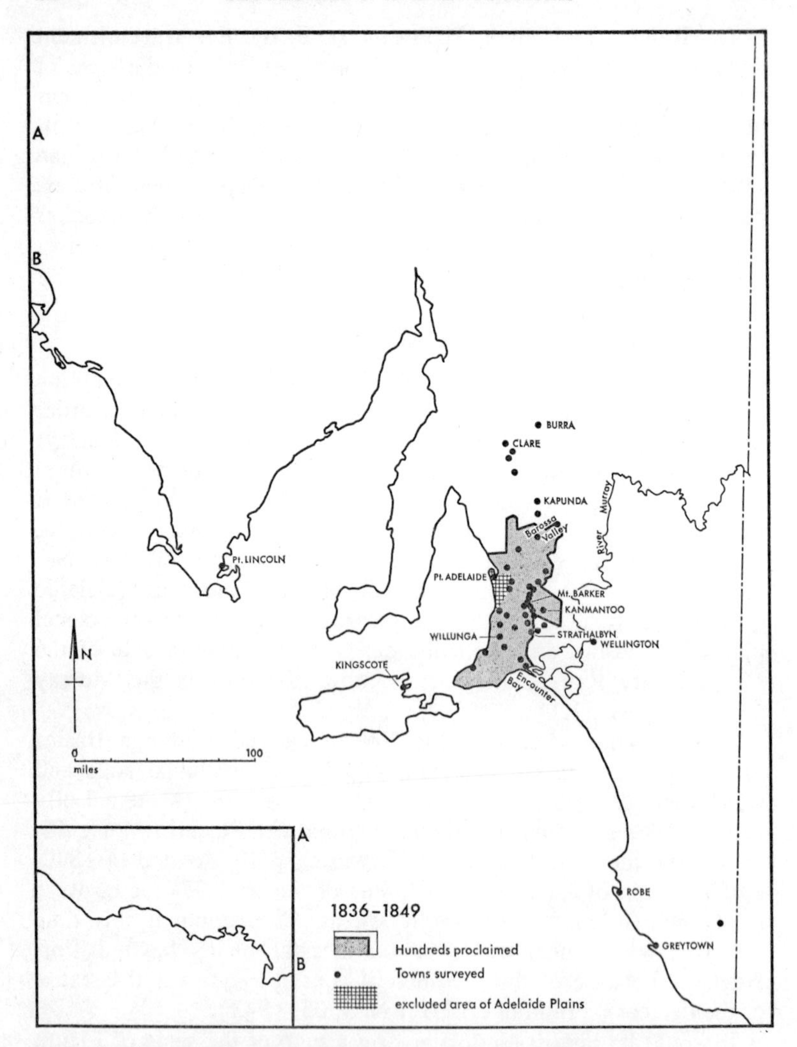


FIG. 5 Spread of settlement in South Australia, 1836-49

few of these villages were very much larger with populations of between 100 and 600 persons; Hindmarsh, Bowden, and Thebarton were already assuming their industrial character, and Port Adelaide had gathered to itself all the ancillary trades, crafts and professions that befitted the main port of the colony. In the basins to the south, the rural settlements were more widely spaced and larger than their counterparts in the Adelaide plain, the size and the range of facilities of Willunga, Noarlunga, McLaren Vale and Morphett Vale being indicative of the agricultural fertility and progress of these areas.

The second concentration of settlement on the eastern side of the ranges was also in a well-watered situation and possessed a savannah woodland cover, but of blue gums (*Eucalyptus leucoxylon*) which, if anything, gave a denser tree growth than that on the coastal plains. Nevertheless, the woodland was open enough to encourage settlement, and by 1840 a string of small hamlets was established near Mt Barker and south towards Strathalbyn, and 2,036 acres of land were enclosed, of which 646 were under cultivation. Although little land was taken up for agriculture between Strathalbyn and the Murray mouth at this time, the possibility of river navigation and trade was a constant lure to settlers in the eastern foothills. Particularly enticing was the area near where the River Murray broke through the barrier of coastal dunes and bars, because a port of transhipment was almost bound to be located somewhere in the vicinity. Already two small settlements existed around the shores of Encounter Bay and Lake Alexandrina and sixty-three acres were in cultivation.

Beyond the main core of settlement in the ranges lay isolated communities like Wellington, a crossing point on the River Murray; Port Lincoln, at the tip of Eyre Peninsula (an alternative but rejected site for the capital); and Kingscote, on Kangaroo Island, a whaling station.

The central Mount Lofty Ranges were avoided because of their steep slopes, podzolized soils, and their cover of *Eucalyptus obliqua* forest.

The later years of the decade saw the consolidation of agriculture in the original areas of settlement and the establishment of isolated pockets of settlement in other parts of the colony; in the Barossa Valley about thirty miles north-east of Adelaide, near Clare in the Central Hill country and at Robe and Greytown in the Lower South-East where the pastoral estates needed shipping points. But of greater significance than all these outposts of settlement was the discovery and exploitation of copper, at Kapunda in 1843 and at Burra in 1844, both of which discoveries focused attention on the Central Hill country, and brought about a major influx of capital and people into the colony. By the end of the period in 1849, minerals constituted about 67 per cent of the value of exports of the infant colony, compared with 29 per cent for wool and 4 per cent for wheat and breadstuffs. These figures indicate the importance of mining in the economy at that time. Yet all this should not lead one to under-estimate the progress that was being made in agriculture. Once the initial period of experimentation, and literally acclimatization by the predominantly British population to their new 'Mediterranean' environment, was over, the arable land was expanded. In 1845 there were 1,267 farmers and 26,218 acres of cultivated land, but by 1848 the number of cultivators had risen to 1,846 and the acreage had nearly doubled to 48,912. The local invention of Ridley's 'stripper', a small-scale harvesting machine, admirably suited to the small 80-acre sections that predominated at this time, was a factor in the expansion of agriculture and a portent of the ability of the South Australian pioneer to adapt himself to, and to overcome his new environment.

The original Wakefieldian ideal of a self-supporting peasant agriculture on freehold farms, worked by a sturdy middle class yeomanry, was all but achieved in these early years. The newly domesticated landscape of the coastal plains and basins and of the eastern slopes of the Mount Lofty Ranges, south to Encounter Bay, was subdivided into a regular grid of 80-acre sections that followed the cardinal points of the compass, and splayed uncompromisingly across rivers and hills, nearly each section a property in itself, with its home and barns. The scenes of activity as the land was 'rescued from a state of nature' were exciting and the cause of some pride and sober self-congratulation, for the South Australian pioneer was well aware of the solid success he had made of his colonization venture which was in marked contrast to the beginnings in other colonies.

## PROSPERITY, 1850-1859

The comfortable and steady progress of the colony in the previous years might well have continued during this decade had not the gold discoveries in Victoria in 1851, with their accompanying increase in population by over 500,000 in the next ten years, opened up a new market for the South Australian farmer.[3] With a rush of optimism new hundreds were proclaimed towards

[3] Pike (1957: 181) suggests that there was a net out-migration of probably about 10,000 persons from South Australia to Victoria in the initial years of the 'diggings', causing some depression of trade, but many of these returned later and started farming with their newly won wealth.

the Murray, but more particularly towards the Central Hill country which had a vegetation cover and soils similar to the already well-tried and known areas on the flanks of the Mount Lofty Ranges. Moreover, it was an area to which attention had already been directed by the copper mines of Burra and Kapunda (see Figure 6).

*Developments, 1850-1854*

On the whole, the optimism was misplaced and land in the new hundreds was not taken up because there was still plenty available in the old nucleus of hundreds near Adelaide. As a consequence, during the first half of the decade, wheat acreages in the Adelaide plains and southern basins increased noticeably. The land was well farmed and well fenced and 'reaping and winnowing machines were everywhere'. Viewed from the adjacent upland slopes the plain and basins with 'their square enclosures of varied crops had a map-like effect' (S.A. Reg., 3 and 10 March 1851). The prosperity of these coastal areas was reflected as much in the towns as in the land itself, and the four existing large centres of the southern basins were added to with the creation of Reynella and Aldinga. Some extension of cultivation also occurred in the valleys which cut into the steep cliff-like edge of the Fleurieu Peninsula: Yankalilla, Rapid Bay and Myponga each had a small amount of land in cultivation, extensive sheep runs in the back country and a jetty to export their produce.

The new demand for food succeeded in breaking down the prejudice that existed about the plains immediately north of Adelaide. Salisbury was in this new area of colonization and was said to have 'sprung up like a mushroom'. But beyond Salisbury and the Para River, which was regarded as a northern frontier of agriculture in the plains, lay the Peachy Belt, a vast stretch of peppermint gums the importance of which, wrote a local correspondent 'for fencing and firewood . . . at no very great distance from Adelaide, is almost incalculable'. It was thought to be 'inexhaustible' but small clearings were already being made on its edges and by about 1880 it had all but disappeared.[4] Beyond Gawler, the fine trees of the Peachy Belt degenerated with decreasing

[4] *The Adelaide Observer*, 30 July 1859, had this to say about the colonists of the Peachy Belt, 'how far they are doing right by waging a war of utter extermination against the timber, I do not pretend to say; but certain it is that unless peace be proclaimed in time, their descendants will look in vain for a tree in that which was once Peachy Belt'.

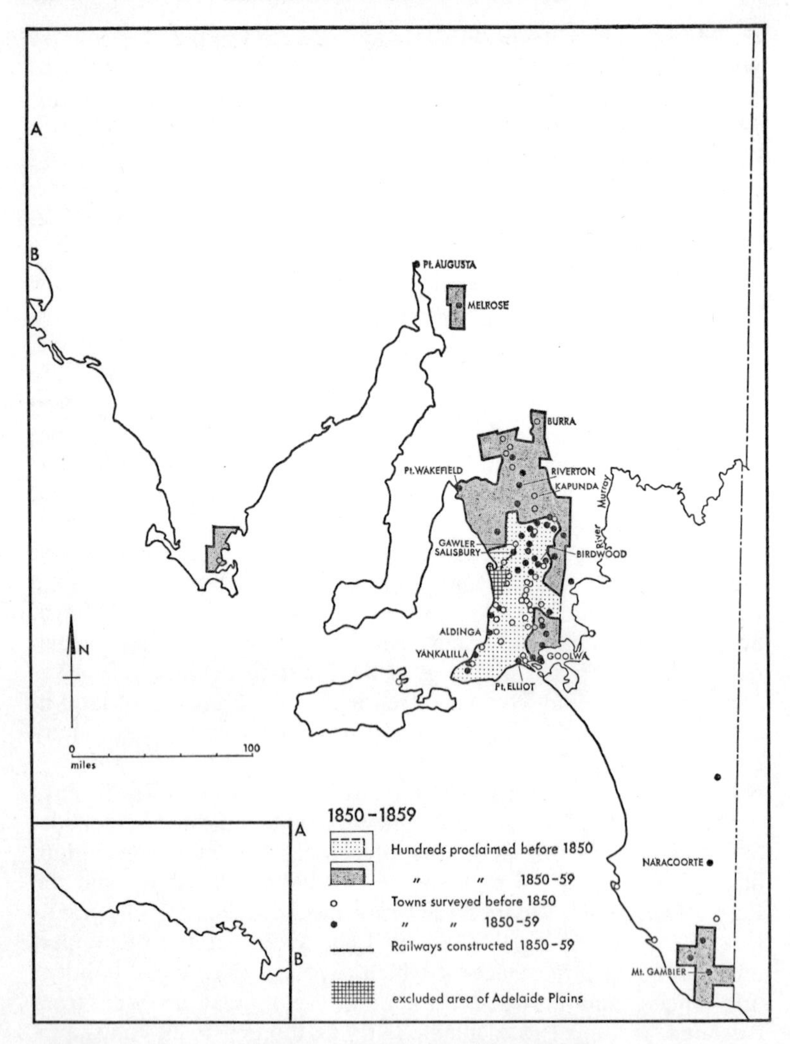


FIG. 6 Spread of settlement in South Australia, 1850-59

rainfall into the Gawler Scrub, a zone of dwarf mallee of 'amazing extent' that covered the coastal plains to Port Wakefield, and which defied clearing and settlement for the next twenty years (Figure 9) despite the fact that it was blocked out into hundreds as early as 1856 (S.A. Reg., 3 and 22 May 1851).

On the eastern side of the Mount Lofty Ranges the greatest expansion of settlement had occurred in the upper valleys of the Rivers Torrens and Onkaparinga where the extent of cultivation, the number of townships and the capacity and number of flour mills were beginning to rival those of the prosperous southern coastal basins. But the great need was for an outlet for the produce of the area and from 1856 onwards farmers throughout the whole of the eastern plains and foothills agitated very strongly for a tramway to Goolwa to join the line that existed to carry the river cargoes landed at Goolwa to the ocean-going ships docked at Port Elliot (S.A.P.P.: No. 205, 1855-56; No. 20, 1857-58; No. 17, 1859; and No. 75, 1865-66). The Barossa Valley was being actively colonized, but north and beyond it there was very little settlement, except for the bustling mining townships of Kapunda and Burra. Yet the lack of farming around these mining settlements did not mean that the landscape was not changing. The Kapunda smelters alone needed 120 tons of wood a day, and the local timber was felled indiscriminately. The Burra smelters consumed 150 tons a day, but there were no supplies of wood in the vicinity as the vegetation changed from open woodland to rolling 'steppes' that were described as being 'an abomination of desolation'. The nearest fuel supplies lay in the mallee scrub several miles to the east and one observer noted that 'drays loaded with wood were arriving every instant . . . to replenish the forest of cut logs and long wood piled up, circling the whole area of the works, and filling every available space' (S.A. Reg., 26 June and 1, 3, 5 and 8 July 1851).

Away from the main body of settlement, agriculture was beginning in a new and separate nucleus, in the fertile, volcanically-derived soils near Mount Gambier in the Lower South-East district. Expansion was limited by the extensive swamps to the west and north, but attention was being paid to the well-drained savannah woodland that flanked the swamps in the east and extended along the border with Victoria towards Naracoorte. Like their counterparts in the eastern foothills of the Mount Lofty Ranges, the farmers of Mount Gambier were agitating for cheap transport to the coast, and investigations into different tramway routes to various ports were carried out in 1858. It was a measure of the success and promise of this new settlement (S.A.P.P.: No. 38, 1859).

New isolated townships arose at Melrose near the Mt Remarkable copper mine in the southern Flinders Ranges and at Port

Augusta which was a major wool outport for the extensive pastoral estates that stretched northwards along the line of the Flinders Ranges and southwards towards the main body of agricultural settlement. Port Wakefield, at the head of Gulf St Vincent, exported Burra copper and later imported coal for the mines and smelters.

*Developments, 1855-1859*

The second half of this decade, however, saw a marked change in the locale of colonization. Up till then the modest expansion of townships and agricultural settlement had occurred in the hundreds proclaimed before 1850. But after 1856, a new and distinctive sub-region of colonization began to emerge in the western part of the Central Hill country, from Kapunda to Clare, in the hundreds proclaimed in 1851. Between 1856 and 1859, 56,000 acres of wheat-land were added to South Australia, of which over half was new land taken up in the new hundreds north of the Barossa Valley in County Light where there had been only 4,000 acres in crop in 1855. Undoubtedly the construction by 1860 of South Australia's first major railway, through the wheatfields of the plains north of Adelaide to Kapunda provided a new outlet for the produce of the emerging region and encouraged settlement.[5]

To sum up, the number of acres of land under cultivation had risen from 64,949 in 1850 to 361,884 in 1854, the number of cultivators had risen from about 2,500 to about 7,000 in the same period and the population had nearly doubled from 63,700 to reach 122,735 by the end of the decade. South Australia was riding high on a new prosperity based on a triad of products, copper, wool and wheat, of which wheat and breadstuffs now accounted for 43 per cent of exports by value, compared with 31 per cent for wool. At the time South Australia was pre-eminent amongst Australian colonies in the amount of land under cultivation. The attainment of self-government in 1856 was one of the obvious manifestations of this new scale of sufficiency and prosperity. With the early struggles to gain a foothold in this new and strange land successfully overcome and behind them, the colonists were soon to take a longer look at the problems of colonizing the less attractive lands of the immense domain that they had inherited.

[5] This was not the first railway in South Australia. Prior to this, in 1856, a line had been constructed between Adelaide and Port Adelaide, whilst a horse-drawn tramway existed between Goolwa on Lake Alexandrina and Port Elliot for the transhipment of cargoes from the River Murray steamers to ocean-going vessels.

Not only did their relations with their own land need organizing, but also their relations with the wider world, where lay the market to which they were tied (Williams, M. 1968).

## TRANSITION, 1860-1868

The northward-probing agricultural frontier hastened the development of the new sub-region as 'an industrious body of farmers' colonized the upper valleys of the Gilbert and Light Rivers and the low, rolling hills between, adding about 100,000 acres of wheat to the colony's total by 1867. The townships of Riverton, Saddleworth, Stockport and Clare had all been established during the preceding period and by 1860 were flourishing rural centres. They were joined by Rhynie (1860), Hamilton (1863), Marrabel (1864), Manoora (1864), Allendale (1864) and Tarlee (1868). A typical example of their growth was Rhynie which was described in 1866 as being a 'rapidly improving place', with the full array of mechanics' shops, tradesmen's shops, a hotel and a flour mill, whereas there had not been a dwelling home 'within miles' some five years before. Similarly, Marrabel was 'only about two and a half years old but in a flourishing and promising state' with ninety inhabitants, a flour mill, two general stores, two blacksmiths and a wheelwright (Whitworth 1866: 130, 139, 192). The new area of colonization was not served by a new railway for the line from Roseworthy was not extended to Tarlee until 1869. Wheat hauling to the ports was done by bullock drays on indifferent and poor roads. In the coastal plains west of the Central Hill country, colonists were cautiously edging into the thin soils and mallee scrub lands, and by gruelling hand-felling nearly 76,000 acres were cleared and put into cultivation between 1860 and 1867, and the new townships of Redbanks and Mallala were created to serve the new farmlands (Figure 7).

A new tier of hundreds was proclaimed to the north after 1863, with an obvious westward extension to the base of Yorke Peninsula where the three new copper mining townships of Wallaroo, Kadina and Moonta had a total population of nearly 8,000 by 1865. Almost without exception the naturally open grasslands in the new northern hundreds from Clare to Burra fell into the hands of the pastoralists who were able to outbid the intending small selectors and so establish themselves on freehold instead of leasehold estates, as was previously the case. The five government townships of Euromina, Canowie, Davies, Anama and Hilltown

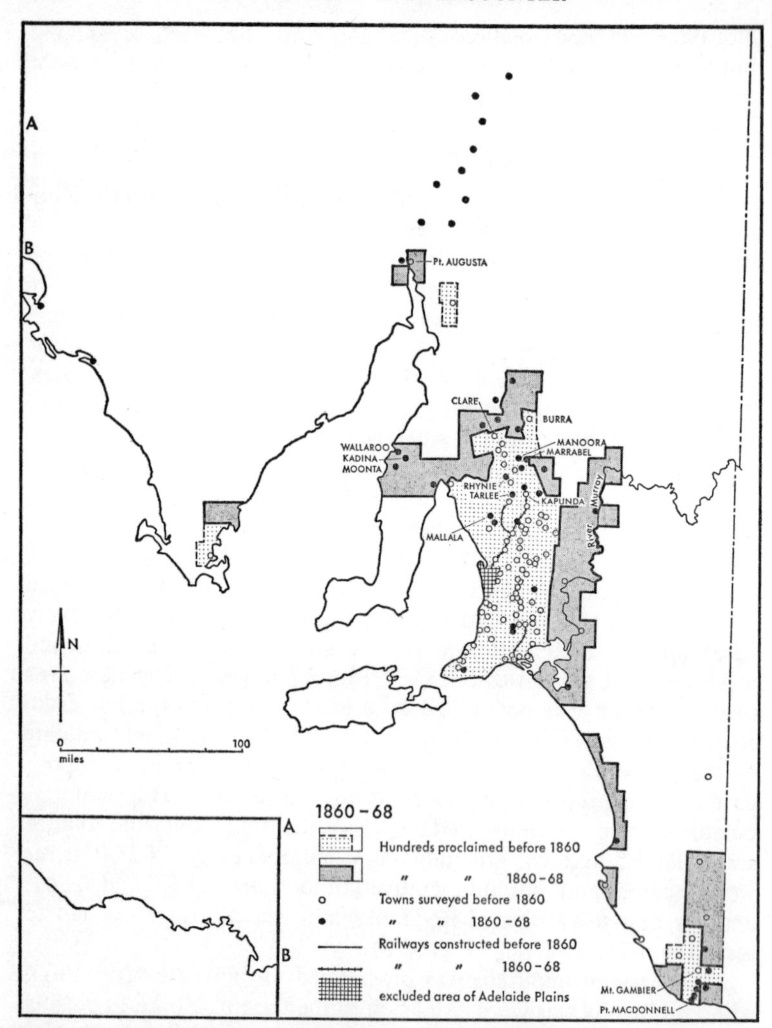


FIG. 7 Spread of settlement in South Australia, 1860-68

surveyed to encourage agricultural settlement in this area were extinguished from the landscape by similar purchases, and a 'cordon of pastoral country' locked up settlement in the north (S.A.P.D.: Col. 1002, 1868-69, and Col. 258, 1865).

On the eastern side of the main body of settlement in the hill country, a belt of twenty-four hundreds was proclaimed in 1860, extending either side of the River Murray from the great North-west Bend to the mouth. These were, for the most part, in impenetrable mallee country and were not indicative of settlement at this time. It transpired that the hundreds had been proclaimed as a result of the personal animosity of the Commissioner of Crown Lands against a pastoralist and in 'killing the squatter' (a popular and accurate phrase for proclaiming hundreds at this time), the Commissioner had resumed all the other land with no regard for the bulk of leases (S.A.P.P.: No. 158, 1860, and S.A.P.D.: Cols. 267-69, 589-602, 1860).[6]

Away from the main body of settlement, two new wool ports were established on the west coast of Eyre Peninsula and six new mining centres were scattered in the Flinders Ranges, beyond Port Augusta. In the South-East district a large block of hundreds was proclaimed in anticipation of agricultural settlement in the woodlands around Mount Gambier and along the border with Victoria, but like the new hundreds extending from Clare to Burra, the bulk of this land was bought by the pastoralists. With the exception of the vicinity of Mount Gambier, where the location of new townships indicates the extension of the thriving but small nucleus of cultivation of approximately 26,000 acres by 1867, virtually no settlement took place in the South-East.

*Difficulties and Decline*

The expansion of colonization began to slow down after 1866 and for the first time the total acreage of land under wheat in South Australia fell, from 550,456 acres in 1867 to 533,035 acres in 1868. The decline in acreage was regarded with alarm because wheat was the basis of prosperity and expansion, accounting for exactly half of South Australia's export earnings. There were many reasons for this hiatus in the progress of settlement, the most specific being drought and the heavy incidence of disease in wheat. It was also realized that farmers were leaving South Australia to settle on the grey and black clays of the Wimmera district of Victoria. But the cause of this migration, exaggeratedly described as 'an exodus', lay not in the drought, but in the difficulty of obtaining new land in South Australia compared with the ease of obtaining it in Victoria (S.A.P.P.: No. 20, 1868-69 and No. 124,

[6] From here on, the process of hundred proclamation was formalized. A map of the proposed hundred or hundreds with detailed descriptions of vegetation and soil characteristics had to be laid before Parliament for approval. Thus from 1860 onwards, the hundred becomes an even more useful guide to the spread of settlement than before.

1867 and S.A.P.D.: Cols. 532, 575 and 659, 1867). The migration was but an obvious manifestation of a more deeply rooted malaise that had set in many years before.

It was difficult and expensive to get new land. The small farmer was battling against the 'long purses' of the pastoralists and against the land agents or 'land sharks' as they were popularly known. The rights of commonage appended to the 80-acre sections, and taken in the unsold land of the hundreds, were now extinguished as all land was becoming alienated. It was impossible to make a living from cultivating an 80-acre section and also provide grazing for draught animals. In addition, the increasing use of mechanical harvesters and seeders necessitated bigger blocks and farms (S.A.P.P.: No. 158, 1860, Qu. 112-17; No. 73, 1865, Qu. 665-66).

The farmers felt hemmed in by the pastoralists to the north and the dense mallee vegetation in the dry, sandy plains to the west and east which extended like 'an uninterrupted waving prairie' and was so thick that if a road was cut through it the mass of trees stood up 'like high walls on each side'; indeed, because of the umbrella-shaped canopies the effect was described as being not unlike 'a road through a tunnel'. To jump beyond the pastoral belt was not viewed with enthusiasm by farmers who were 'afraid to go beyond this into the open country for fear of the bush fires and the isolated position of their farms' (Whitworth 1866: 134 and S.A.P.P.: No. 73, 1865, Appendix). Above all, transport was the problem, as anything that increased the price of wheat in the face of Californian and Chilean competition was of little interest. To go into the nearer mallee land was almost unthinkable under existing conditions of tenure. It had to be cleared laboriously by hand-felling and the stumps grubbed out, and those that remained suckered again after rain. Water for stock was totally absent and the minimum offering price of £1 per acre was not likely to prove attractive to the would-be scrub clearer.

Hence agitation arose to open up the scrub lands that touched upon the fringe of the well settled lands, by granting easy and liberal terms to those who were disposed to clear them (Figure 8). In 1866, the Scrub Land Act was passed setting aside Scrub Districts in which land was to be offered for sale as usual, but if unsold after one month was to be divided into blocks of up to 640 acres each and leased for twenty-one years with right of purchase at the minimum upset price of £1 per acre, provided that the holder cleared one-twentieth of the holding each year (for their location see Figure 8). In this way it was hoped 'to change the

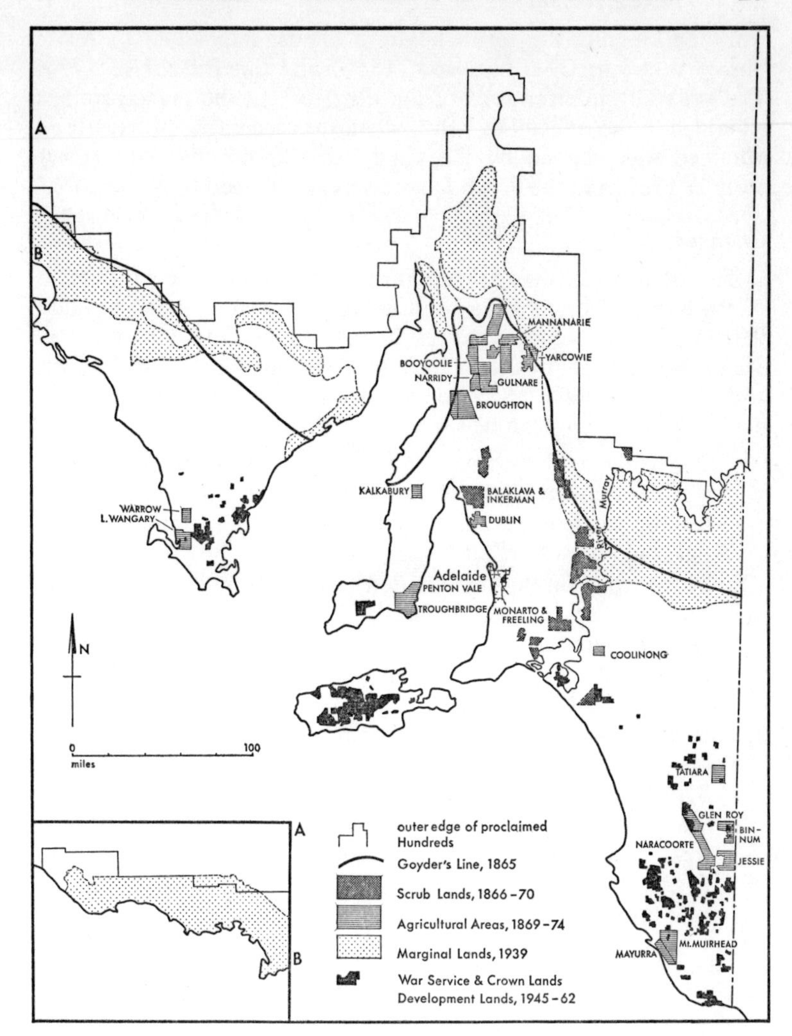


FIG. 8 Five aspects of South Australian land settlement

face of a large area of the country from hopeless scrub to smiling fertility'. But all in all, the limited experiment was not a success as selectors found difficulty in complying with the regulations, particularly as many of the Scrub Districts were situated a few miles distant from the existing frontier, thus making part-time or evening work on clearing the scrub holdings almost impossible (S.A.A.: Sur.-Gen. Com. O/L No. 463, 1866, and S.A.P.P.: No. 172). The generally inferior soils of the districts and the mallee stumps all added to the difficulties.[7] But, whatever success might have been obtained was vitiated by the passing of Strangways' Act (q.v.) early in 1869, and the Scrub Districts almost passed from memory.[8]

*Changes*

The problem of the depression of 1867-68, and the experiment of the Scrub Districts have been looked at in some detail because they marked profound changes in matters of land settlement in South Australia that were to have far-reaching geographical effects and were to mark a transition from one distinctive phase to the next. These changes can be summarized as follows:

1. The Wakefieldian assumption that all land was of equal value was demolished with the creation of the Scrub Districts, and a more realistic appraisal of the colony's agricultural potential could now be made.
2. There was formal recognition of the long-held view that survey sections larger than the usual 80 acres were needed, as settlement was moving to drier areas, areas of inferior soil and difficult vegetation, and because of increased mechanization.
3. Selected limited areas in which an experiment in land settlement could take place were set aside. This became a typical South Australian expedient in the face of new changes.
4. The Scrub Land Act brought about the introduction of what was, virtually, a system of credit sales. Indeed, the Scrub Districts were the thin edge of the wedge that the Wakefieldian supporters feared they would be.

But the innovation and experiment did not end there. In the case of the creation and design of townships, the government

[7] S.A.P.P.: No. 172, 1868-69 is a report which reviews the progress of settlement in the Scrub Areas of Balaklava, Inkerman, Monarto and Freeling. Costs of clearing were usually about £2 per acre, but could be as high as £4 or £5 where the mallee scrub was particularly dense.

[8] It is quite clear from the debate on the Scrub Land Revision Act of 1877 that the details of the legislation of ten years before were completely forgotten. However, the 1877 Act allowed a greater amount of land to be leased, the amount to be cleared annually was reduced to one-fortieth, and land could be purchased at £1 an acre at the end of eleven years. The more generous provisions were merely a reflection of the inadequacies of the first Scrub Land Act.

stipulated after 1865 that at least one township was to be provided for every hundred, in order to aid and encourage agricultural settlement, particularly the establishment of religious, educational and social institutions. So thoroughly was this policy pursued that there was little opportunity left from here on for private speculative creations, and this policy does much to explain the fairly dense but even cover of township sites in Figures 9 to 12. As with land settlement, so with transport. The government had attempted to sell its interest in the Port Adelaide and Kapunda railways to private operators in 1865, but could not get a good enough price. Realizing, however, the fundamental value of the railway in developing new farming regions and getting the harvest to the ports, the government made construction and operation its sole prerogative (Williams, M. (1966b, and S.A.P.P.: Nos. 22 and 98, 1865, Nos. 20, 20A, 20B and 141, 1865-66).

In every sense this was a period of transition in South Australia. The purely theoretical concepts laid down in another country, halfway around the world, had worked fairly well up till now, but were totally inappropriate for rapidly expanding colonization, as the accelerated pace of settlement in the Central Hill country had revealed. To be true to the ideals of stability and order in the colonization process in new areas and under new conditions, the government had to make its control of settlement more flexible, but paradoxically it also had to make it more widespread because the basic character of colonization was changing. Gone was the ideal of the self-supporting yeomanry and in its place there was emerging a new, specialized agriculture of wheat growing for a worldwide, competitive market. New land, and lots of it, was needed for this new kind of agriculture and the principle of 'concentration' was receding in the face of rapid expansion. In the basically underdeveloped economy of the colony, the government was probably the only body capable of providing the capital and organization for the works needed. In any case, the utilitarian philosophy was deeply rooted and accepted in South Australian thinking, and most persons naturally looked to the government to promote the greatest good for the majority of the population, and ensure order in the face of the potential chaos of colonization; while the government accepted, naturally, the role required of it. All this is not to say that the government had no pre-conceived notions of what was needed, but unlike the rigid adherence to Wakefieldian principles in the past, the notions were modified according to circumstances. Pragmatism took the place of theory.

## EXPANSION, 1869-1879

The expansion of colonization during the next eleven years was spectacular. The legislation responsible for this 'unreasoning boom' (Roberts 1932: 398) was the Waste Lands Amendment Act of 1869, or Strangways' Act as it was popularly known, which for the first time allowed land to be bought on credit in specially selected localities considered suitable for cultivation and called 'Agricultural Areas'. Born out of the difficulties of the depression of 1867-68 and the limited experiments of the Scrub Leases, Strangways' Act substituted credit sales of large blocks of land up to 320 acres in extent—the successful bidder paying 20 per cent down and the balance in four years—for the old system of purchasing 80-acre sections for cash. Certain safeguards were taken to preclude the speculative land agent and the pastoralist by making the settler prove his *bona fides* by occupying his land within six months and residing on it until his purchase was completed.

Armed with this new promise for success the practically-minded and experienced South Australian farmers of the old settled areas, who had been struggling to gain a living from small sections and small farms, stepped over the constricting belt of sheepwalks and moved into the Northern Areas; looked south to the promised land of the Mount Gambier district; and cast a hesitant glance at the little known and isolated tip of Eyre Peninsula. A new phase in the colonization of South Australia had been initiated.[9]

### *The Agricultural Areas, 1869-1872*

The aggregate mass of hundreds proclaimed and townships surveyed during this remarkable phase and shown in Figure 9, masks many locally important sub-movements. Obviously the first expansion occurred in and near the Agricultural Areas in the Northern Areas (for their location see Figure 8); the immediate and main locale of colonization activity was in a zone bounded approximately by the townships of Port Pirie, Port Broughton, Yacka and Jamestown (the Broughton, Gulnare, Narridy, Yarcowie and Mannanarie Agricultural Areas for example). The initial progress was slow, but with an excellent harvest in 1870-71 and yields of between twenty-three and thirty bushels to the acre, the settlement of this zone was accelerated and land bought eagerly. A second frontier

[9] Except where specific quotations indicate otherwise, or where reference is made to the colonization of Eyre Peninsula, or the South-East district, this section is based on Meinig (1962: 28-77).

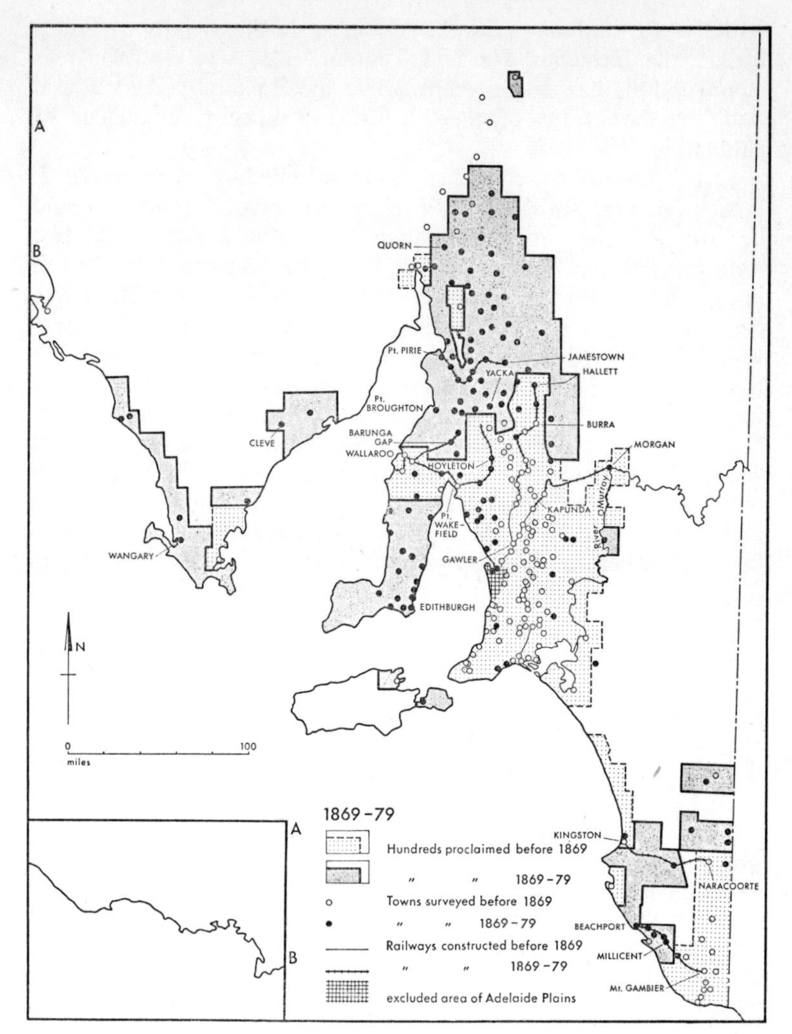


FIG. 9 Spread of settlement in South Australia, 1869-79

of colonization was located on the edges of the mallee-covered plains between Gawler and Port Wakefield and settlers moving north from Adelaide and westwards from the hills were undoubtedly encouraged by the credit lands of the Dublin Agricultural Area, and by the Port Wakefield–Hoyleton horse-drawn tramway completed in 1870. In the heel of Yorke Peninsula, in the vicinity of Edithburgh, the Troughbridge and Penton Vale Agricultural Areas stimulated similar development, while the Kalkabury Agricultural Area in the centre of the Peninsula was another nucleus of cultivation.

In the South-East, completely different circumstances prevailed. It was felt that the delays in draining the coastal swamps would be overcome by proclaiming them Agricultural Areas and two were created, the Mount Muirhead and the Mayurra. But the full implication of this decision was not realized immediately. It was one of the regulations of Strangways' Act that land within the Areas would be offered for sale at a maximum price, and that if it remained unsold the price would be reduced at roughly yearly intervals until the minimum price of £1 per acre was reached. Settlers were not going to take up flooded and water-logged land at high prices but would wait until the prices fell. If the government was committed to drain the land for agricultural purposes at a price ranging between 16s. and £2 per acre, then the land had to be drained before the third year of offering, at least, in order to recoup the minimum land revenue and the expenditure on drainage. Therefore the government unwittingly committed itself to extensive and immediate draining activity. The land was withheld from sale and 67 miles of drains were cut in the flats, and gaps cut through the coastal dunes and ranges that impede the egress of flood water. By 1872 the land was opened for credit selection, though still not thoroughly drained (Williams, M. 1964).

In the border areas with Victoria, the Naracoorte Agricultural Area was created. Settlers, particularly from the eastern foothills of the Mount Lofty Ranges, came to this Agricultural Area only to find that some of the land was under water during the winter and that the regulations about *bona fide* settlers had been flouted flagrantly by local pastoralists who had bought up the land to fortify their leasehold estates. Many settlers 'trekked' into Victoria where free selection was in force, whereas a few camped in the district in the hope that the other Agricultural Areas of Lawson, Jessie and Binnum would be opened soon (*Border Watch*, 2 Feb., 26 March and 6 April 1870).

### *Goyder's Line and Beyond, 1872-1879*

Although the wheat acreage had increased by about 43 per cent, from 499,937 acres in 1869 to 715,776 acres in 1872, the poor season of 1872 had resulted in a mere 29 per cent increase in the entire yield during the same period. The farmers were disgruntled. In the South-East the demand was for more draining, more and bigger Agricultural Areas, and a more stringent suppression of bogus selectors. In the Northern Areas the farmers complained of the government's restriction of credit sales to the Agricultural Areas and demanded that all land south of Goyder's Line, the line demarcating the northward limit of 'safe' agriculture, be opened to them (Figure 8) (Meinig 1961). This was approved in 1872, and was accompanied by a reduction of the deposit to 10 per cent, the extension of credit to six years and the enactment of provisions about cultivation to dissuade speculators and pastoralists.

With good rains and good yields during the next two years settlement surged northward again. Townships blossomed in and around the Agricultural Areas and another quarter of a million acres of land were in cultivation by 1874. The undoubted success was a heady wine and the farmers were intoxicated with the prospect of even greater prosperity and demanded 'that Mr. Goyder's rainfall [Line] be shifted out of the colony' and that all land in South Australia be opened to credit sales. Eventually the government came round to their view, and in a debate in the Assembly, Blythe, the Chief Secretary, spoke in favour of the farmers' desire to explode the artificial barriers to settlement.

> It was a little singular that the question of what was pastoral and what was agricultural land had been gradually extending year after year. He was old enough a colonist to know the time when it was asserted that land north of the Para River was not fit for cultivation, and could not be ploughed. Subsequently agricultural settlement was extended to Gawler, and they had gone on step by step until a feeling was entertained two years ago that the line of rainfall laid down by the Surveyor-General for defining the classes amongst the pastoral lessees, might be fairly taken as the limit of agricultural land. . . . It was generally admitted that this line of rainfall was one which should not be allowed to continue to exist as an obstacle to further settlement. As an obstacle it needed to be removed.
>
> (Meinig 1962: 52, and S.A.P.D.: Cols. 905-6, 1874).

Thus, between 1874 and 1879 a whole new sub-region of colonization came into being in the northern and eastern edges of the main body of settlement around the Agricultural Areas as settlers moved into the inter-montane basins of the Flinders Ranges and reached the mining outposts established twenty years before, even reaching the edges of the seemingly endless saltbush plains that stretched away into the interior. As Goyder's Line was passed and agriculture safely established beyond, the old idea 'that the rain follows the plough' was evoked to explain the undoubted 'change of season'.

Local optimism ran high, probably nowhere more so than in the many new townships surveyed on the advancing northern frontier. A description by a country correspondent, of one, Quorn, must suffice for the description of other successful settlements, and the unfulfilled aspirations of many more. 'This township has now commenced in earnest and, instead of kangaroos, we have now masons, carpenters, stone carriers and labourers hard at work. It will be a race between Mr. Greenslades' hotel and Mr. Armstrong's which will be up first. We have two stores already in active operation and one boarding house, one butcher and baker and temperance dining rooms in course of erection, and five or six private homes and two flour mills' (*Port Augusta Dispatch*, 27 July 1878).

At the same time, the difficult mallee-covered plains of Yorke Peninsula and the eastern side of Gulf St Vincent were the scene of active clearing and colonization which now progressed with a new vigour with the advent of two important inventions. First, Mullens, a farmer from near Wasleys, north of Gawler, discovered in 1868 that the slender mallee trunks could be knocked down by dragging a heavy roller over the ground, and that after firing the trash a tolerable crop could be obtained by dragging a spiked log over the uneven surface of roots and stumps, followed by sowing the seed, broadcast. The technique of 'mullenizing', though crude, was effective, and in time became adopted widely; but the invention of the 'stump-jump' plough, accredited to R. B. Smith of Kalkabury, Yorke Peninsula, in 1876, completed the conquest of the mallee lands, for its hinged shares rose out of the ground if they hit the large and particularly solid mallee roots that mullenizing left behind (Callaghan and Millington 1956: 319-20; Dunsdorfs 1956: 155-57; Shann 1930: 221-22). The 'despised' scrub lands now became attractive, the old Scrub Land Act was revived in 1877, and 600,000 acres of scrub land were taken up in the next four years.[10]

One of the greatest obstacles to the spread of settlement had been overcome by the experimental ingenuity of practical farmers. It

[10] Meinig (1962: 105) points out that under the new regulations of 1877, greatly enlarged acreages could be leased in the Scrub Lands, the amount of land requiring clearing was reduced to one-fortieth and the land could be purchased for £1 per acre at the end of eleven years.

was not without a wider significance either, for these inventions and later innovations and discoveries were to show that the practical farmer often had the answer to successful colonization, not the government which merely laid down the framework for settlement once it was occurring. It also goes a long way towards explaining why the government was so ready to relinquish its role as sole arbiter in land matters during the 1870s. In many instances theoretical planning could not achieve as good a result as practical experiment.

*Communications*

All these solid achievements in the field of land settlement were not made without improved means of communication. The success of the 1872-74 thrust northward had emphasized the inadequacy of transport facilities from farm to ports, and of the ports themselves to handle the abundant harvest. Communications became a matter of great and consuming concern. The extended and indented coastline of South Australia encouraged the establishment of a multitude of small ports and nearly all entertained hopes of becoming the focus of a regional railway system. Clearly only a few were going to succeed. The success of Port Wakefield, with its short line to Hoyleton, was repeated at Wallaroo with its twin lines, one towards Barunga Gap (constructed 1878-79) and one to Port Wakefield itself which ultimately robbed the latter of much trade because Wallaroo was a far superior port. Port Pirie's inland line tapped perhaps the richest wheatgrowing land of all, and by 1878 it had reached Jamestown. Adelaide, although far away, was not left out in this race for trade; the new branch from the Kapunda line reached Burra by 1870 and Hallett by 1878, so tapping the eastern fringes of the newly settled areas and constituting a potential threat to Port Pirie's trade. The line to Morgan on the River Murray was not built in anticipation of any rural settlement, so much as to capture the river traffic from Victoria and New South Wales and direct it to Port Adelaide. This dealt a cruel blow to the ports around the Murray mouth.

In the South-Eastern district, events from 1872 to 1878 had not gone so well and the region had lost in the race of improvement compared with the Northern Areas. The construction of the line from the coast at Kingston (1876) to the Agricultural Areas near Naracoorte was premature because of the failure to establish a thriving agricultural community in the border lands in the face of pastoral competition. But perhaps a worse mistake than this was the fact that the construction of the line encouraged the opening up, between 1876 and 1878, of a compact block of six hundreds in the inland swamps. Generally speaking, these hundreds were in the worst drained portion of the South-East and it was absurd for the government to offer them for sale. Realizing the difficulties that had occurred in the coastal swamps when land was sold before drainage was completed, the Commissioner of Crown Lands decided in 1879 to reserve from sale any lands liable to flooding, pending an inquiry on how best to drain them. Severe flooding throughout the flats of the South-East in the same year confirmed the wisdom of the decision. More successful initially was the Beachport–Mt Gambier line (1879) which linked the rising Millicent area and the well established agriculture zone of Mount Gambier with the coast (Williams, M. 1964).

The third region of colonization activity was on Eyre Peninsula, but the Lake Wangary and Warrow Agricultural Areas were near failures. Elsewhere in the peninsula lack of rail communication tied settlement to within 10 or 15 miles of the coast, consequently the potentially better soils inland were not touched. Except for the nucleus of settlement on the isolated patch of relatively fertile red-brown loams near Cleve and Cowell the sparse settlement of the region was confined mainly to the high rainfall western coast of the peninsula, and it is significant that the whole region was for long known as 'the West Coast' and not Eyre Peninsula.

## STAGNATION AND EXPERIMENTATION, 1880-1892

The opening of the 1880 season showed all the signs of promise that the previous year had led the settlers to expect and a whole new tier of townships and hundreds was surveyed and proclaimed by the government in 1880 on the eastern and northern fringes of the settled area. But the promise did not hold; by the end of the season drought conditions were widespread, yields dropped to below two bushels per acre in those very hundreds created in the previous year and the total South Australian yield fell from 14·2 million bushels in 1879-80 to 8·6 million bushels, to fall again by another million bushels during the next two years. The situation in the Quorn district was perhaps typical of the persistent optimism in the face of adversity that characterized the northward thrusting frontier fringe. In the same article, the editor of the *Port Augusta*

*Dispatch* talked about the growing prosperity of the town, its shops which made the main street look like a 'miniature Rundle Street' (the main retail thoroughfare in Adelaide), and yet 'every week brings news of disappointed hopes . . . as to the late harvest. The farmers are lamenting the past and dreading the future' (7 January 1881).

One bad season should not have been so disastrous, but as Meinig points out, the general success of the previous years had obscured the grim fact that only marginal crops were being obtained on this frontier, and the 1880-81 drought was not, 'as elsewhere, an abrupt set-back, it was a harsh prolongation of defeat' (Meinig 1962: 79). The next year, settlers' attention shifted towards land near the railway that was being constructed from Terowie to Orroroo and Quorn, and which promised reduced marketing costs. Also, the higher rainfall, but neglected, mallee lands north of Wallaroo were looked at with new respect, and a group of towns was surveyed in this new area of colonization (Figure 10).

At the same time, many pioneers stayed on in the fringes of settlement in the hope of recouping some of the investment in land and labour, but the continuing drought caused them to despair finally and to ask the government for some relief, either in the form of seed for the next season or the revision of credit repayments. But the government refused. It was a strange situation and one that had its roots firmly in the past heritage of theory and planning. The government had always set itself up as the arbiter in land settlement matters, but had gradually relinquished control at the behest of the farmers themselves, who, in the past ten years, had claimed that they were better judges of what was good for them than was the government. Yet who was to blame? The farmers who wanted to break down the preconceived theories of colonization, or the government which let them? It was a nice argument, but like similar disputes over the draining of the swamps in the Lower South-East, the government realized that the sooner something was done to alleviate the situation, the easier and cheaper it might be. Thus relief was provided for those drought-stricken settlers who got an average yield of only six bushels per acre or less during the preceding season, together with a right to surrender their holdings and re-select their land. After a continuation of the drought in 1882-83 these relief provisions were extended to allow settlers to convert their credit purchases into long-term leases.

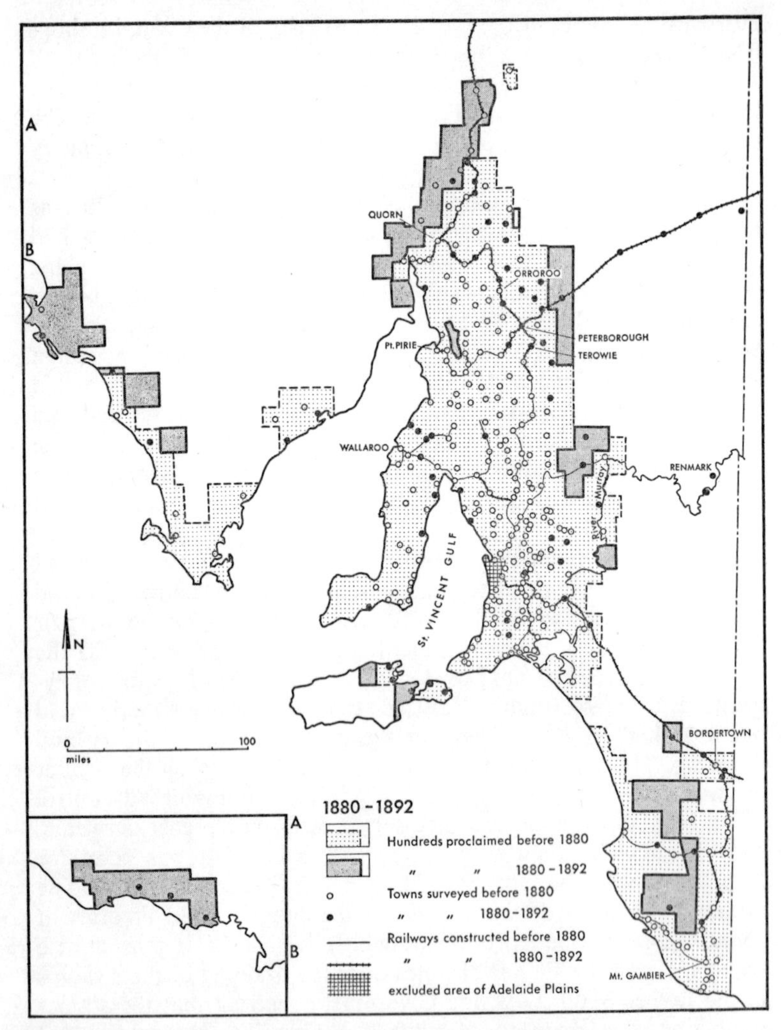


FIG. 10 Spread of settlement in South Australia, 1880-92

The droughts of 1880-83 were a set-back to settlement that finally prescribed the approximate limits of agriculture in the northern and eastern parts of the State. Many of the townships surveyed during this time never progressed beyond a pattern of surveyor's stakes in the ground, or at best, they were permanently stunted communities, parodies of the grandiose plans from which they started. Even with better seasons after 1884 there was little desire to move out again into the saltbush plains. Consolidation occurred in the 'safe' areas for agriculture, the population immediately north and east of Port Pirie increasing by over 4,000 between 1881 and 1891, but rarely extending too far beyond the northern railway line. This tended to mark the margin of the saltbush country and the probable limit of successful cultivation.

The South-East was largely untouched by the drought, but any spectacular increase in agricultural settlement was completely strangled by the 'cold embrace' of the pastoralists. The only successful agricultural endeavour was on the deep clays of the Bordertown district through which the Melbourne to Adelaide railway had been constructed between 1881 and 1887. The acreage under wheat surpassed that of the Mt Gambier district by 1883, but the wheat production of the whole of the South-East was trifling compared with that of the Northern Areas, which were less humid and more suitable. Plantings of wheat for the South-East fell from a peak of 36,238 acres in 1890 to 21,571 acres in 1892. The only advance made in the settlement of the Lower South-East was the partial drainage of the inland swamps, which is reflected in the distribution of new hundreds (Figure 10). The partial drainage was achieved by opening up natural watercourses that trended in a north-west to south-east direction between the low ridges and flats that form the characteristic topography of the South-East. The channels were improved from south to north and hundreds were blocked out and proclaimed between 1883 and 1888 from south to north as some of the surface water was cleared away. But the drainage works conferred very little lasting benefit because the outfalls were nearly 200 miles away from the places they were intended to drain, and the fall was so gradual (less than two feet per mile) that barely a movement of water was achieved, let alone the rapid evacuation of flood water from the high rainfall areas of the south.

Elsewhere, tentative advances were made in the eastern end of Kangaroo Island, and the peripheral settlement of Eyre Peninsula was augmented in the far west by the creation of some townships and the opening up of new land when pastoral leases fell in during 1889 and 1900, and the land was reserved for agriculture. Compared with the other outlying region of settlement, the South-East, the increase in wheat acreage in Eyre Peninsula was moderately impressive and more than doubled between 1880 and 1892 when it reached 46,154 acres, while the population increased by 1,800 during the same period. Optimism now ran high and one member of the Assembly optimistically foresaw the west coast of the peninsula as 'the California of Australia'; capable of supporting 'hundreds of thousands of families' (S.A.P.D.: Col. 1815, 1887).

But all in all, the droughts of 1881 had set the tone for the rest of the decade and South Australia entered a prolonged period of depression and agricultural stagnation which was reflected in the decline in land under cultivation from 2·7 million acres in 1884 to 2·5 million acres in 1892. Surprisingly enough, the hardest hit areas were not those on the fringes, but in the core of the old settled area in the Central Hill country where the population dropped by 4,300 and the wheat acreage fell by 164,000 acres.

*Experiments*

A few lights pierced the economic gloom. First, there was the successful establishment of an irrigation settlement at Renmark on the River Murray by the Chaffey brothers of California in 1888, which pointed the way to developments in the future, and secondly, there was the discovery in 1884 of the Broken Hill silver-lead-zinc mine which stimulated the building of the railway from the inner edge of the wheat-lands at Peterborough to Broken Hill, and the establishment of smelters at Port Pirie. Less obvious at the time, but far more important for South Australia, and indeed the whole continent in the future, were four discoveries and experiments that were to provide solutions to the colonization problems of some of South Australia's difficult lands.

First, there was the growing realization that the old methods of rapacious cultivation were reducing yields, and that superphosphate manure was needed to stabilize the down trend in yields. J. D. Custance, the first principal of Roseworthy Agricultural College near Gawler, pioneered the use of superphosphate in 1881 and his successor, William Lowrie, forcefully advocated its use. The techniques of its application, with seed, to the field were worked out by farmers from near Minlaton on Yorke Peninsula in 1892, and by the end of the century over a quarter of the farmland of South Australia was fertilized and yields were not only stabilized but began to rise (Callaghan and Millington 1956: 91-93, 143-55; Cornish 1949).

Secondly, South Australian farmers were experimenting with fallowing as a means of conserving moisture, controlling weed infestation and preparing the seed-bed, before scientific research refined the technique.[11]

Thirdly, from 1880 onwards, experiments with drought and disease resistant varieties of wheat like Steinwedel and Early Gluyas, to mention only two, were developed by farmers from Dalkey and Port Germein, and these varieties were widely used and contributed to the extension of wheat growing into drier areas. These selections preceded the cross-bred varieties of Farrer (Callaghan and Millington 1956: Chaps. XVI and XVIII, and Dunsdorfs 1956: 189-92).

Finally, there was the discovery in 1889 by A. W. Howard, a Mt Barker nurseryman, of the potential value of the introduced plant, subterranean clover, which was to be found occasionally in the Adelaide Hills but had previously been considered insignificant. He unceasingly extolled the drought resisting and soil nitrogen-fixing qualities of the plant, made the first harvest of its seed (a major technical accomplishment because it is buried), and after many years of publicizing, established it as a recognized sown pasture legume. The impact of this and other varieties and other legumes was not great at the time; it was not until after 1920 that the plants were widely accepted. But today subterranean clover is the most important of all sown plants in Australia, including wheat, and it has transformed the appearance, economy, and yield of the old mono-culture wheat regions, and has been one of the tools in the reclamation and settlement of the moderately and very difficult sandy lands (Donald 1958).

It was indeed a period of folk experiment and innovation that first began and was elaborated in South Australia. If these four methods and techniques of colonizing new lands, and raising fertility and yields in existing settled lands, are added to the developments made in machinery to cope with the mallee lands in previous years, then the South Australian contribution to agriculture in Australia is conspicuous.

## RAILWAYS AND NEW LANDS, 1893-1920

The depression of the 1880s continued into the next decade and was accelerated by severe droughts in 1896 and again in 1901. A virtual halt occurred to all expansion, attention was turned pri-

[11] W. Campbell of Dakota is credited with pioneering the technique of 'dry farming' in the 1890s, but Lowrie's remarks before the First Interstate Dry Farming Conference in 1911 suggest that some South Australian farmers were practising it by 1888. See Callaghan and Millington (1956: 115).

marily towards the Central Hill country and only 70 miles of new railway were constructed and six new townships created between 1893 and 1902. In fact, a definite contraction of settlement was under way as the droughts were stiff reminders to those settlers who had ventured to the very fringes of the northern wheat-land between 1880 and 1892 (Figure 11) that their livelihood was precarious and about 3,000 persons left the northern and eastern fringes by 1902. Everywhere, continual cropping without fertilizing was taking its toll and the average wheat yield which had been steadily decreasing over the last twenty years plunged to 1·66 bushels per acre in 1897-98, the lowest it had ever been, thence to rise again only with the increasing application of superphosphate fertilizer and the practice of fallowing.

Not only did conditions get no better in South Australia at this time, but the attractiveness of other States increased. The Western Australian gold fields at Kalgoorlie opened in 1897, and in the years that followed, the State's very liberal land laws which were tantamount to free selection, attracted many struggling South Australian farmers. In Victoria, settlers were edging successfully into the fringes of the mallee country at the beginning of the 1890s, and similarly this had caused a marked migration of farmers across the border from South Australia (Roberts 1924: 330 and 298, V.P.P.: No. 1, 1891).

With the turn of the century, however, the demand arose that the government open up new land in order to assist expansion. The slowly rising confidence was occasioned by the good results of superphosphates, better strains of wheat, and the slow but steady acceptance of the technique of 'dry farming', all of which coincided with a slight increase in seasonal rainfall, which was explained as a result and not a cause of the success of the new techniques (Dunsdorfs 1956: 204). The only land left was in Eyre Peninsula and the Murray Mallee, both difficult areas with light, sandy soils but supposedly amenable to dry farming, both with a dense cover of mallee eucalypts and both well away from the coast. For settlement to be successful in these areas the farmer needed to be primarily a wheat grower and not a wheat carter, and in order to promote settlement the government reasserted its authority by building an extensive network of 'developmental railways', so initiating a second great movement of pioneers comparable in scale, and sometimes in substance, to the 1869-80 surge into the Northern Areas.

With understandable caution, considering the past experience, it was decided to concentrate attention on the 'Pinnaroo Lands' in

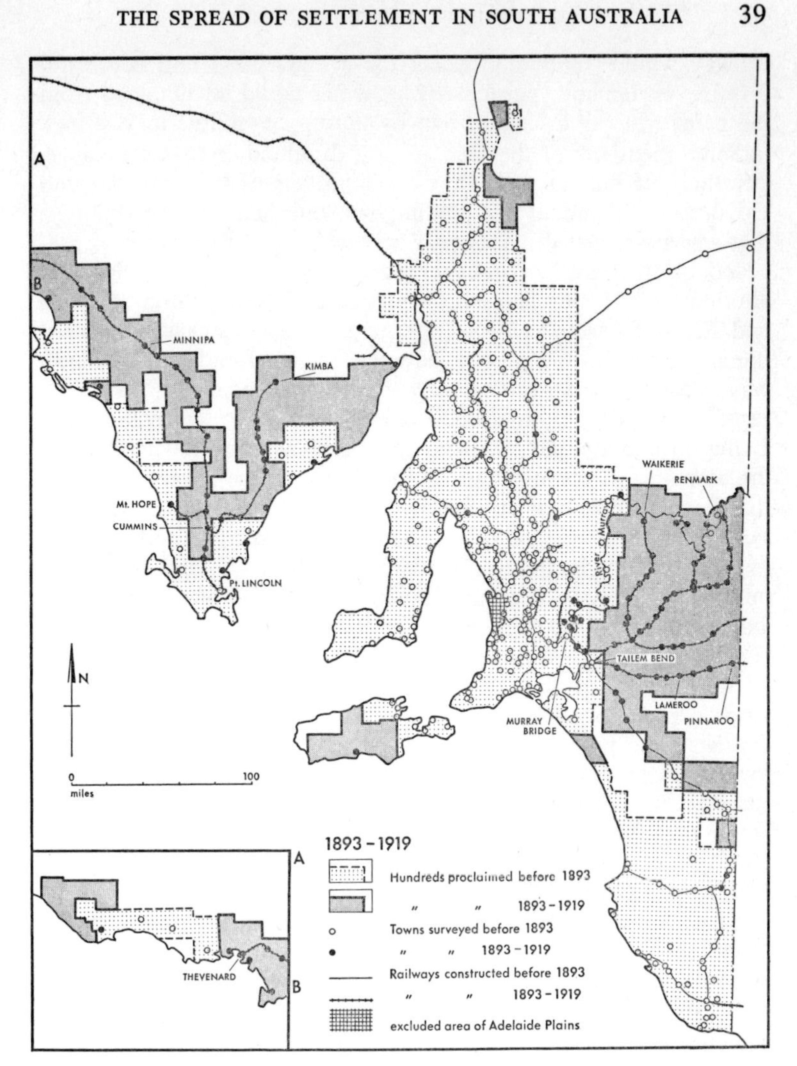


FIG. 11 Spread of settlement in South Australia, 1893-1919

the south-east corner of the Murray Mallee, where Goyder had previously located about a quarter of a million acres of better grade soil that would improve with the application of superphosphate but would still require extensive scrub clearing. A Commission of Inquiry in 1902 felt that the provision of a railway would 'offer a prospect for a new, large and successful farming settlement', particularly as water could be obtained from the great Murray artesian basin by boring down 150 to 200 feet.

Some members of the Commission dissented from the findings, pointing out that in their view the land was near, if not 'beyond Goyder's Line' and all that that implied. But the Pinnaroo Railways Act was passed in the next year and teams of surveyors who preceded the railway in 1904 marked out 146,000 acres which was allotted to 161 purchasers. These figures rose to 246,000 acres and 226 purchasers in 1905. The pulse of settlement was quickening so that when the railway between Tailem Bend and Pinnaroo was opened in 1906 settlement was said to be occurring 'at a very rapid rate'; 40,000 to 50,000 acres were under cultivation and Lameroo and Pinnaroo were emerging as thriving towns to serve the new area of colonization. In the next year the government sent the Surveyor-General to the United States to inquire into the system of dry farming and he returned impressed and convinced of its applicability in South Australia. By 1908 all land within 10 miles of the railway was taken up and the success of the settlement 'had been almost phenomenal'. The Surveyor-General commented, 'The whole area, which a few years ago only kept about a dozen persons and a few head of stock, and was considered by most as a worthless desert, is now one of our most prosperous farming districts. The population is estimated at 2,700, several flourishing towns are in existence, last year 825,409 bush. of wheat were reaped, and 6,669 acres of crop were cut for hay; a very much larger area is now under crop and there were in the district about 3,000 horses and 1,500 cattle' (S.A.P.P.: No. 22, 1902; No. 78, 1903; No. 10, 1905; No. 10, 1906; No. 10, 1908).

The success of the Pinnaroo venture was not confined to the original area alone, because it focused attention on the whole of the mallee lands as a possible locality for colonization. Indeed, from 1906 onwards settlers had been moving from the irrigation settlements along the River Murray, south and east into the Mallee, particularly in the vicinity of the Big or Pyap Bend, north of Loxton. Concomitant with these developments went the reclamation, by embanking, of the flats that bordered the lower River Murray at, for example, Murray Bridge, Mypolonga and Monteith. Each settlement had a distinctive morphology and the farms were strung out in a long line on the upland flood-free edge of the river, like so many polder villages.

By 1908 the success of the Pinnaroo settlement seemed assured.

The total area under wheat was 155,000 acres and would have been more had the pioneers not been caught up in a vicious circle of events by not being able to pay the high price asked for horse feed and therefore not being able to keep enough horses to clear the ground to grow the feed. By 1910 the acreage had nearly doubled to 317,000 of wheat and the Murray Mallee had passed the Upper North as a wheat-producing area.

The 'second settlement' in Eyre Peninsula was similarly a success and the land which formerly had 'hardly a hoof on it' became a flourishing wheat region. For years the peninsula had been looked upon as 'being tied up for the squatter and it was only recently discovered that there was practically a province for South Australia to settle her people' (S.A.P.D.: 1905, 108). With the construction of the Port Lincoln to Cummins railway in 1907, the whole of the southern part of the peninsula was blocked out into hundreds and alienated, so that by 1910, 367,000 acres of wheat were under cultivation. Many of the new farmers here, as in the Murray Mallee, came from the northern and eastern fringes of the Upper North, which lost about 5,300 people through migration during the first decade of the century.

The northern advance in both new areas of colonization continued; between 1910 and 1916 the survey teams could barely keep pace with the farmers' demands. 'With the use of phosphates and other manures and a scientific system of farming,' said Strawbridge, the Surveyor-General, 'hardly any class of land can be said to be valueless for agricultural purposes.' Indeed, the dry farming technique with its 'long fallow', a ploughing in January or February, and the maintenance of a weed-free mulch until seeding in the autumn of the following year, seemed to hold the key to Mallee settlement. The procedure was claimed to conserve the whole winter's rainfall so that the following crop would have the support of two years' rain, and some enthusiasts even suggested a fallow extending over two winters, so that theoretically at least, three years' rainfall would be available to the crop (S.A.P.P.: No. 10, 1910; Callaghan and Millington 1956: 118).

Goyder's Line had long since been passed in the Murray Mallee, while it was fast being approached in Eyre Peninsula, not that many knew or cared where it was. Certainly the results were impressive and whereas there were only three settlers between the Murray River and the Victorian border in 1904, in 1911 there were 'two large towns, and six or seven smaller ones, whilst settlers' homes (the population is approximately 3,900), bores and other improvements are visible in all directions throughout the district and the clearing of the land is constantly being extended'. In language strongly reminiscent of about forty years before, the Premier told Parliament in 1912 that 'a new chapter in the history of South Australia was opening out, and they were glad to know that their country was being regarded as a bigger State than they had ever previously imagined it to be'. The opening of the Tailem Bend to Renmark railway in 1913 and the two other branches of the railway line to Loxton and Waikerie in the Mallee in the next year, together with the extension of the Eyre Peninsula lines to Kimba, Minnipa and Mount Hope in 1913, seemed to be practical confirmation that the State was bigger than ever before (S.A.P.P.: No. 10, 1912; S.A.P.D.: 1912, 1325).

The declared policy of leaving no wheat farm more than 15 miles from a railway, the coast, or the River Murray was pursued methodically and without exception, and does much to explain the digitate pattern of railways. The provision of sidings at approximately five-mile intervals also gave a distribution of potential township sites that was as distinctive as it was generous. Not all could grow, for in these wheatgrowing areas the potential hinterland population of any one township was not much beyond ten or twenty families, and time was going to have its effect in sorting out the most rewarding sites, something which can be observed at the present time.

But the unthinking expansion in the Mallee and Eyre Peninsula beyond the well-tried limits of Goyder's Line, was bound to come to grief one day, and the drought of 1914 told with melancholy effect on the new areas. In the Mallee, no area recorded a yield of more than half a bushel per acre and 'absolute failure' was common. The position in the north half of Eyre Peninsula was the same. But unlike the great reversals of the 1880s one bad season was not enough to absorb the vapour of optimism, and with the great assistance provided by the government in the form of seed, manure and fodder, it was expected that 'a very much larger area of land will be planted for cereals than has been cultivated for that purpose in the past'. The next harvest was indifferent and settlers were cautiously advised to diversify and put more land down to grass, keep sheep, pigs and poultry, though official opinion was unabated that a 'thriving population of agriculturalists would soon appear'. But the 1916-17 season did nothing to improve matters, and although over 1,000,000 acres were given over to wheat in the Mallee and Eyre Peninsula, those who had remained on the land from two years before were now in 'a hopeless condition'. Were it not for government assistance, there would have been 'a wholesale abandonment of the country' (S.A.P.P.: No. 10, 1915; No. 10, 1916; No. 10, 1917). Indeed, retreat was in evidence, and with the Surveyor-General's admission that perhaps the Murray Mallee was 'one of the least favoured portions of the agricultural areas', debts mounted, rents were reduced, and it was declared in 1920 that 'the state had erred in the past in regard to the land settlement policy in the outside country' (Roberts 1924: 314).

There was a limit to what superphosphates, new wheat varieties, dry farming and new railways could do and these developments affected the basic limitations of the wheat-land soils between 1910 and 1920. Drought assistance and the emergencies of war had pushed and encouraged settlement too far. Instead of stabilizing the position, it had worsened it. In every way the optimism of twenty years before was checked.

## 1920 ONWARDS

The last forty-five years is the lengthiest and perhaps the most complex of all the phases to analyse. Except for the extension of the Wanbi–Yinkanie line in the Mallee in 1925, and the Wandana–Penong and Kimba–Buckleboo lines in Eyre Peninsula in 1924 and 1926, there was no new outward movement of settlement (Figure 12). The outer limits had been set and recognized, and from now on the blanks in the existing pattern of hundreds and townships were going to be filled in, for example, along existing lines in Eyre Peninsula. But more than that, it was a period of intense reorganization of settlement within the existing framework, and that makes trends difficult to analyse.

### *The Marginal Lands*

The outer Mallee areas were still claiming attention as new zones of colonization in the early twenties but conditions were definitely deteriorating. Attempts to recoup the cost of clearing, combined with the need to burn the stubble to discourage the mallee regrowth meant a few years of almost continuous cropping. In addition, the dry mulch theory called for frequent cultivation so that overcropping and frequent tillage meant that soil fertility and structure were damaged. Crop failures and very low yields became common, but it was soil exhaustion and wind erosion (particularly after the 1929 drought) that ruined the farmers, and not the droughts which made symptoms acute and dramatic. With

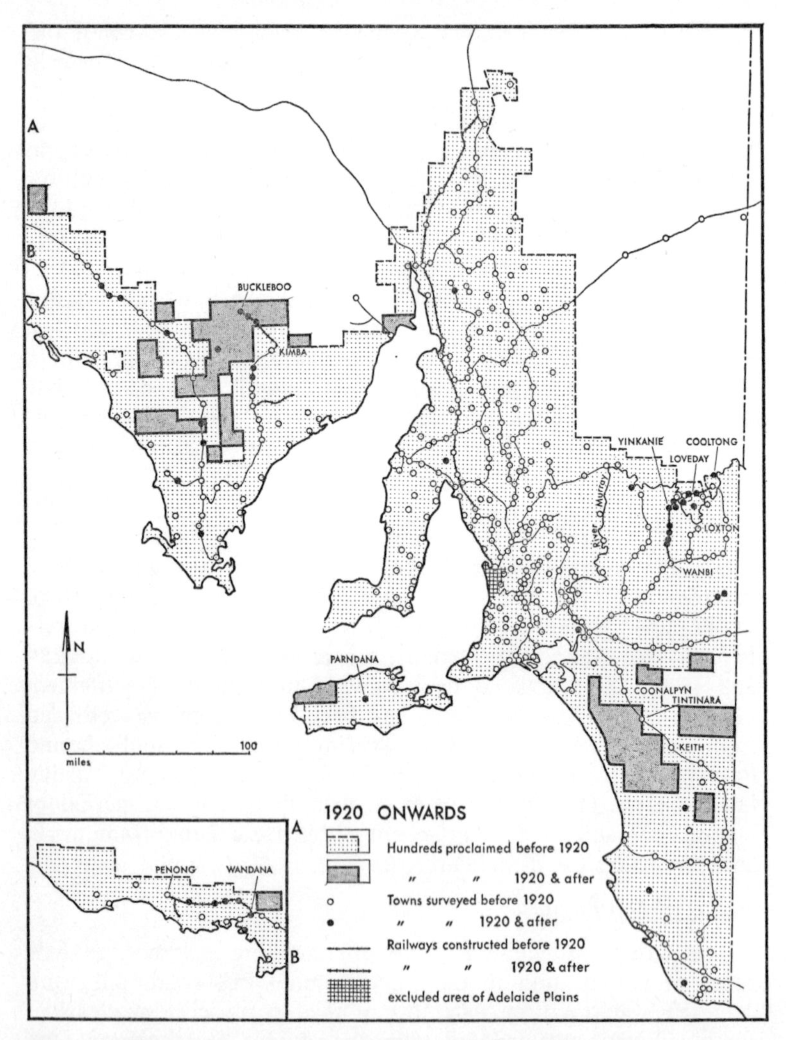


FIG. 12 Spread of settlement in South Australia, 1920 onwards

the collapse of the wheat market in the early thirties it was realized that there was a zone of land that originally had been subdivided into blocks intended primarily for wheat farming, but which, owing to inadequate rainfall and unsuitable soil, had proved unsuitable for mono-culture wheat and should be diversified with livestock grazing (S.A.P.P.: No. 62, 1939; No. 52, 1947). Hence the more realistic concept of the 'marginal lands' (Figure 8), which had long been in the minds of thoughtful people, came into being and replaced Goyder's Line as a 'frontier' of settlement. Throughout the 1930s the problem of the rehabilitation of the marginal lands was discussed but action foundered from lack of funds.

In 1939 the Marginal Lands Committee was formed, and with Commonwealth Government backing of £½ million per annum, it set about the task of reducing the 1,276 settlers in the 1,695,000 acres of marginal lands to about 659 by amalgamating properties, making the existing units more usable by aiding in the clearing of mallee scrub, erecting stock fences and providing water supplies. By 1947, most of the desired programme had been achieved and a large measure of stability returned to the marginal lands. Since that time, the voluntary process of amalgamation of properties has continued and with the widespread introduction of legumes and livestock, which replaced fallow, the process of change has been accelerated so that a grain/sheep farming complex has come into being. Up to 50 per cent of the population has left most of the marginal lands and empty homes stand in the fields, mute testimony to this migration. These changes, together with the provision of better roads and the high percentage of car ownership in Australia, have combined to rob the local townships of what little custom they had and many are on the verge of complete extinction as functional units. Indeed, the marginal lands are characterized as much by marginal towns as by marginal farms.

*Trace-Element Deficiencies*

The discovery after about 1945 that a variety of seemingly infertile soils in the Upper South-East, the tips of the Fleurieu, Yorke and Eyre Peninsulas and the western half of Kangaroo Island could be made productive with the addition of trace-elements such as zinc, copper, molybdenum, and potassium, as well as the usual phosphorus, led to major progress of settlement in these areas (Riceman 1945, 1950). When this discovery was coupled with the introduction of subterranean clover, and other nitrogen-fixing legumes, the results in the settled areas can only be described as revolutionary.

In particular, some of the land in the Upper South-East acquired new significance when it was surveyed into hundreds and sold, one private insurance company pioneering the venture as an investment, and creating some 200 farms on 400,000 acres. As the landscape altered the original name of the district, the Ninety-Mile Desert, was changed to the more prosaic Coonalpyn Downs. No new towns were created, but the original ones like Keith, Tintinara and Coonalpyn became more important and extended their range of facilities.[12]

*Soldier Settlement*

Special conditions for the settlement of returned soldiers were not confined to the end of the Second World War—they had occurred after the First—but the scale of the settlement is significant and large areas of land have been involved under government sponsored and operated schemes (Figure 8). By 1965, 620,149 acres of mainly waste land were reclaimed by clearing scrub and correcting trace-element deficiencies, while some of it was provided by the subdivision and reallocation of previous large estates. Eighty properties have been created in southern Eyre Peninsula, 164 in Kangaroo Island, sixty-six in the Upper South-East and 367 in the Lower South-East where, in addition to the normal process of land reclamation by trace elements, extensive drainage work has been carried out. Major east-west tending drains have been cut across the region, breaking through the barrier of longitudinal low ranges that impeded the egress of water to the coast, and a dense network of minor artificial channels drains the swamps. This post-war settlement and resettlement has not resulted in the creation of many new towns, only Parndana in the centre of Kangaroo Island, and three other centres in the Lower South-East, the latter not growing owing to competition from existing centres. In addition to the 'dry lands' settlements, 312 new irrigation holdings have been created along the River Murray, mainly at existing centres like Loxton, Loveday, and Cooltong near Renmark (S.A. Depart. Lands).

## CONCLUSION

It is the successive stages of settlement that form the principal subject of the present study, and the towns and hundreds enable one to give a detailed and accurate dating and an absolute chronology to the rapid changes in the spatial extent of settlement. It is a method which, with slight modification, could be applied to other States of Australia and so enable one to build up a picture

[12] Details of private land development are difficult to find, but two small booklets do contain much relevant information, i.e. M. C. Butterfield (1958) and Anon. (1959).

of the spread of settlement throughout the whole of the continent, and provide a framework of known fact on which further studies could be based.

Two subsidiary, though by no means less important themes, run through this study. First, the concept of initially favourable and unfavourable areas for settlement, and secondly, the story of the development of techniques of resource-conversion and change.

The first is a useful tool with which to investigate and understand the broad primary movements of immigrant peoples in a new environment, as in this particular example of South Australia. Nor should its wider implications be lost sight of, for it makes more geographically credible hypotheses concerning early colonizations based on archaeological and historical material. Although supporting Ackerman's contention that in constructing historical models with known data we 'provide an insight into the effects of cultural processes today' (Ackerman 1958: 12), it could be added that we also provide an insight into the cultural processes of the past for which less accurate information is available but in which the historical geographer is often vitally interested in his search for the origins of phenomena.

But the environmental stimuli and barriers to settlement are only a beginning, and tell but half the truth about the spread of settlement. Initial assessments of the environment changed with the development of new methods and machines, and new economic conditions, and the less attractive areas of South Australia certainly did not remain so for long. Looking back, one such reassessment of the environment occurred in the 1860s, before the great surge northward, when mullenizing and the stump-jump plough were perfected and the railway system extended. The decades of the 1880s and 1890s were another period of reassessment and experiment, with fertilizers, new wheat breeds, new legumes and dry farming techniques, which preceded the thrust into the mallee lands. This sequence of pause, reassessment and thrust is almost a recurrent pattern in the spread of settlement in South Australia. Sometimes optimism over the efficiency and success of a new technique went too far and produced some of those peculiarly fascinating (almost peculiarly Australian) stories of spreading and then contracting settlement. In the initial situation, the new settler usually chose the location that required the least expenditure of his very limited resources of money and effort, but ultimately the environment was only as forbidding as the level of technology permitted. It was precisely in this way, with the repetition of trial and error, of experiment and mistake, that the South Australian farmers 'worked in right directions' and finally 'conquered the great difficulties of a new and unknown environment'.

## BIBLIOGRAPHICAL NOTE

Throughout this study, statistics of land in cultivation and the value and amount of exports and imports are based on *S.A. Stat. Reg.*; wheat acreages on E. and L. Dunsdorfs (1956); and population changes on C. Fenner (1929).

## BIBLIOGRAPHY

### *Books and Journals*

ACKERMAN, E. A. (1958). *Geography as a Fundamental Research Discipline*. Department of Geography Research Paper No. 53, University of Chicago.

ANON (1959). *Desert Conquest: The A.M.P. Society's Land Scheme.*

BACKHOUSE, J. (1843). *A Narrative of a Visit to the Australian Colonies.* Hamilton Adams and Co., London.

BAUER, F. H. (1959). *Historical Geographical Survey of Part of Northern Australia.* Part I. Introduction and the Eastern Gulf Region. C.S.I.R.O. Division of Land Research and Regional Survey, Canberra.

BENNETT, J. F. (1842). *Historical and Descriptive Account of South Australia.* Smith, Elder and Co., London.

BULL, J. W. (1884). *Early Experiences of Colonial Life in South Australia.* Advertiser, Chronicle and Express Office, Adelaide.

BUTTERFIELD, M. C. (1958). *The A.M.P. Land Development Scheme.*

BUXTON, G. L. (1967). *The Riverina, 1861-1891: An Australian Regional Study.* Melbourne University Press, Melbourne.

CALLAGHAN, A. R. AND MILLINGTON, A. J. (1956). *The Wheat Industry in Australia.* Angus and Robertson, Sydney.

CLELAND, J. B. (1928). 'The Original Vegetation of the Adelaide Plains', *South Australian Naturalist,* **10**: 1-6.

COMMONWEALTH BUREAU OF METEOROLOGY (1950). *Results of Rainfall Observations made in South Australia and the Northern Territory, 1839-1950.* Department of the Interior, Melbourne.

CORNISH, E. A. (1949). 'Yield Trends in the Wheat Belt of South Australia during 1896-1941', *Aust. Journ. Scientific Res., Series B, Biological Sciences,* **2**: 83-137.

DARBY, H. C. (1962). 'Historical Geography' in *Approaches to History: A Symposium*, ed. by H. P. R. Finberg. Routledge and Kegan Paul, London.

DONALD, C. M. (1958). 'The Pastures of South Australia' in *Introducing South Australia*, ed. by R. J. Best. Melbourne University Press, Melbourne.

DUNSDORFS, E. (1956). *The Australian Wheat Growing Industry, 1788-1948.* Melbourne University Press, Melbourne.

DUNSDORFS, E. AND L. (1956). *Historical Statistics of the Australian wheat-growing industry: Acreage and average yield in counties and divisions: New South Wales, Victoria, South Australia and Western Australia, 1792-1950.* Melbourne University, roneoed, Melbourne.

FENNER, C. (1929). 'A Geographical Enquiry into the Growth, Distribution and Movement of Population in South Australia, 1836-1927', *Trans. Roy. Soc. S. Aust.*, **53**: 80-145.

FENNER, C. (1931). *South Australia: A Geographical Study.* Whitcombe and Tombs, Melbourne and Sydney.

HEATHCOTE, R. L. (1965). *Back of Bourke: A Study of Land Appraisal and Settlement in Semi-arid Australia.* Melbourne University Press, Melbourne.

KENYON, A. S. (1914 and 1915). 'The Story of the Mallee', *Victorian Historical Magazine.* Part I, Discovery and Exploration, **4** (1914) 23-74; Part II, Settlement—The Squatters, **4** (1915) 121-50; Part III, Settlement—The Agricultural Occupation, **4** (1915) 175-200.

LOBECK, A. K., GENTILLI, J. AND FAIRBRIDGE, R. W. (1951). *Physiographic Diagram of Australia.* Geographical Press, Columbia University, New York.

MEINIG, D. W. (1961). 'Goyder's Line of Rainfall: The Role of a Geographic Concept in South Australian Land Policy and Agricultural Settlement', *Agricultural History*, **35**: 207-14.

MEINIG, D. W. (1962). *On the Margins of the Good Earth: The South Australian Wheat Frontier, 1869-1884.* Second Monograph of the Association of American Geographers. Rand McNally and Company, Chicago.

NORTHCOTE, K. H. (1960). *Atlas of Australian Soils, Sheet I and Booklet of Explanatory Data for Sheet I, Port Augusta–Adelaide–Hamilton Area.* C.S.I.R.O., Melbourne.

PIKE, D. (1957). *Paradise of Dissent: South Australia, 1829-1857.* Longmans, Green and Co., Melbourne.

PRICE, A. G. (1924). *The Foundation and Settlement of South Australia, 1829-1845.* Preece, Adelaide.

RICEMAN, D. S. (1945). 'Mineral Deficiency in Plants in the Soils of the Ninety-Mile Desert in South Australia', Part I, *Journ. Coun. Scientific Industrial Research Aust.*, **18**: 336-48; Part II, *Coun. Scientific Industrial Research Aust., Bull.*, No. 234.

RICEMAN, D. S. (1950). 'Mineral Deficiency and Pasture Establishment in the Coonalpyn Downs, South Australia', *Journ. Dept. Agric. S.A.*, **5**: 132-40.

ROBERTS, S. H. (1924). *History of Australian Land Settlement, 1788-1920.* Melbourne University Press, Melbourne.

ROBERTS, S. H. (1932). 'The History of the Pioneer Fringes in Australia' in *Pioneer Settlement*, ed. by J. L. G. Joerg. American Geographical Society, Special Publication No. 14, New York.

SHANN, E. (1930). *Economic History of Australia.* Cambridge University Press, Cambridge.

WHITWORTH, R. P. (1866). *Bailliere's South Australian Gazeteer and Road Guide.* Bailliere, Adelaide.

WILLIAMS, M. (1964). 'The Historical Geography of an Artificial Drainage System: The Lower South-East of South Australia', *Aust. Geog. Studies,* **2**: 87-102.

WILLIAMS, M. (1966A). 'Delimiting the Spread of Settlement: An Examination of Evidence in South Australia', *Economic Geog.*, **42**: 336-55.

WILLIAMS, M. (1966B). 'The Parkland Towns of Australia and New Zealand', *Geog. Rev.*, **56**: 67-89.

WILLIAMS, M. (1968). 'Two Studies in the Historical Geography of South Australia' in *Studies in Australian Geography*, ed. by G. H. Dury and M. Logan. Heinemann Educational Books, Melbourne.

WILLIAMS, R. T. (1955). 'Vegetation Regions', map and commentary in *Atlas of Australian Resources.* Department of National Development, Canberra.

WOOD, J. G. (1937). *The Vegetation of South Australia.* Government Printer, Adelaide.

## *Newspapers and Official Series*

*Adelaide Observer.*
*Border Watch* (Mt Gambier).
*Port Augusta Dispatch.*
S.A. Reg. = South Australian Register.

B.P.P. = *British Parliamentary Papers.*
S.A.A. = *South Australian Archives.*
S.A.P.D. = *South Australian Parliamentary Debates.*
S.A.P.P. = *South Australian Parliamentary Papers.*
S.A. Stat. Reg. = *South Australian Statistical Register.*
S.A. Dept. Lands (1946-1962) = *Progress in Land Development (Non-Irrigation).* Land Development Branch, Dept. of Lands, Government Printer, South Australia.
V.P.P. = *Victorian Parliamentary Papers.*

F. H. BAUER

# Climate and Man in North-western Queensland

*'The struggle to develop the north Australian monsoonal regions of winter drought is one of the most stirring epics in the history of efforts by white races to develop the tropics.'*

A. GRENFELL PRICE *White Settlers in the Tropics*

ONE OF THE most obvious characteristics of the rural Australian is his awareness of his environment; he is as surely its product as are kangaroos, kookaburras and koala bears. Indeed, most Australians who have experienced the full range of their continent's climatic vicissitudes would, if the term were explained to them, admit to being environmentalists, and would probably show a strong streak of fatalism into the bargain. The fact is that in much of Australia the physical environment, particularly the climate, plays a greater role in European man's activities than it does in most other parts of the world where he has established himself. Realizing the dangers I run of slipping into geographic heresy, I would like to draw upon nearly a decade's travel, residence and research in northern Australia to tread the middle ground between environmental determinism and possibilism to show some of the effects that a markedly seasonal climate may have on European settlement.

## GEOGRAPHY, HISTORY AND LOCATION

Australia comes honestly, if unfortunately, by her reputation as 'the dry continent'. Her climatological plight began with whatever agencies positioned much of the continent within the influence of the southern hemisphere mid-latitude high pressure system and well out of the effective tracks of extra-tropical cyclones. This bit of geographical bad luck was compounded when epochs of weathering and erosion reduced her highlands to substantial hills which, while scenically beautiful, are woefully inefficient in coaxing orographic precipitation out of the atmosphere. As a result, 38 per cent of the continent receives less than 10 inches of precipitation annually, and 68 per cent receives less than 20 inches.

Chance brought Australia's first white settlers to somewhat familiar climatic environments on the eastern and south-eastern margins of the continent. When settlement began to push west and north, however, they encountered entirely new conditions in the form of markedly seasonal tropical, arid and semi-arid climates. Explorer, miner, grazier and farmer alike did not understand these environments, and in large measure the white man in Australia is just as entangled with the weather and climate as he ever was. Indeed, he has adapted his activities to the environment. As a result very substantial areas, particularly in the northern and central portions of the continent, have proved to be an embarrassment to a young nation whose people are adventurous but few, and whose financial resources are modest.

From time to time Australians become afflicted with what almost amounts to a national fixation—that something must be done to establish, beyond any question, the white man's claim to these vast areas. Such a 'spasm' is now very much in evidence. Thus far the results have been well short of spectacular, but certainly the most useful is a growing body of solid, scientific information about these heretofore imperfectly known areas. An important part of this knowledge is the record of the white man's experience in these regions; certainly the attempts, successes and failures made by those who have lived and are living in these areas are worthy of consideration. This has led me to use certain aspects of the discovery and settlement of North-west Queensland as a case study.

North-west Queensland is in most physical respects typical of many parts of arid and semi-arid Australia. It comprises some 115,000 square miles of moderately varied terrain lying south of the Gulf of Carpentaria, centred on the junction between the 20th parallel of south latitude and the 140th meridian of east longitude (Figure 13). Topographically this region is unremarkable. About a quarter of it, around Mt Isa and Cloncurry, is rough, rocky and hilly terrain of moderate elevation which sometimes has substantial local relief; the highest points do not exceed 1,700 feet. Surrounding this broken country are even to rolling plains, down and tablelands lying no more than 900 feet above sea level; their principal characteristic is topographic monotony. There is nothing topographically or climatically unique about North-west Queensland; similar country and climatic regimes occur in other parts of the continent and in various parts of the semi-arid and arid world. Its fascination lies rather in the story of human experience in the

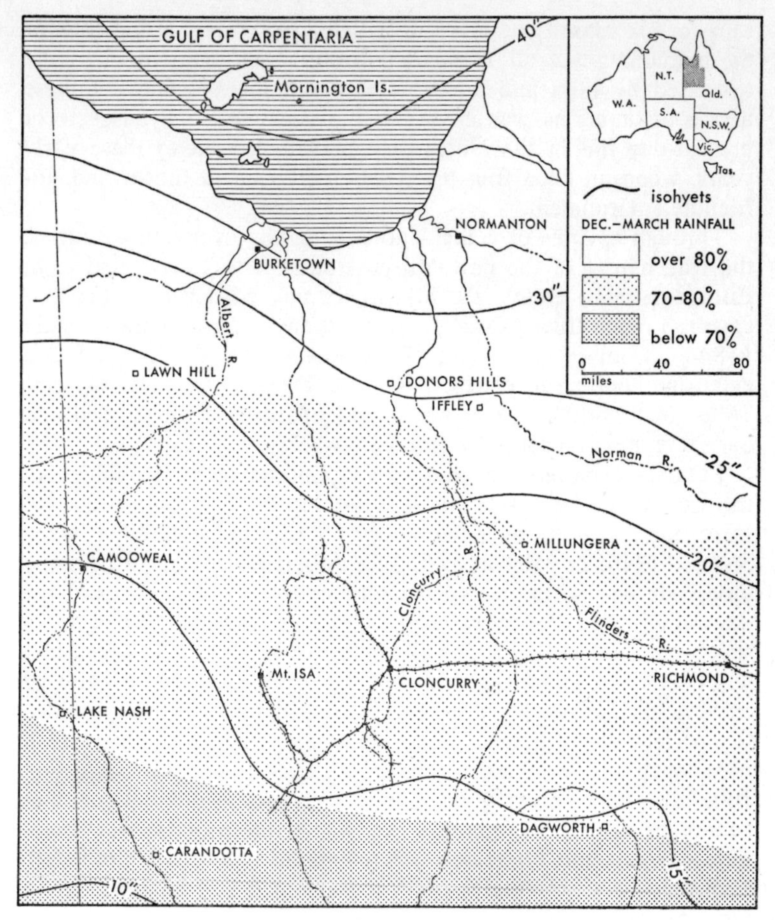


FIG. 13 Settlement in North-western Queensland

region, for it was in North-west Queensland that Australians of European descent first made permanent settlement in the northern part of their continent.

## THE FIRST CONFRONTATIONS

European man came late to Australia and he came with well-established ideas of what constitutes a reasonable climate for the white man, ideas ingrained by centuries of development in a humid mesothermal climate. He knew nothing of a land where it forgets to rain for six months or more at a time, yet where during four of the remaining six all movement on the land may be severely restricted by bogs and floods. The psychological shock of these first confrontations was a very real thing, and many aspects of present-day life in the North-west may be traced to these early years, when an alien folk first encountered an unknown and unfriendly environment.

Through a series of coincidences extending over four centuries, the true nature of the northern environment was concealed from Europeans. The Dutch, the first to see any portion of it (1623), gave it a much truer character than did any explorer until Gregory (1848). Coming at the end of the wet season, the Dutch found extensive flooding in places, but in the end Carstenszoon (Heeres 1899: 34) said: '. . . it is our judgment this is the most arid and barren region that could be found anywhere on the earth.'

Flinders explored the Gulf of Carpentaria at the beginning of the wet season of 1802, but saw nothing of the countryside. No European had been more than a few miles inland until 1841, when Stokes (1846, 2: 319) ascended the Albert River some 50 (river) miles. He was so impressed with the country that he foresaw the day when '. . . the now level horizon would be broken by a succession of tapering spires rising from the many Christian hamlets that must ultimately stud this country.' As a final gesture he emblazoned across the then empty map a name which provided the perfect bait, for what land seeker or politician could resist a name such as 'The Plains of Promise'? Seeing the Gulf country in June and July, before grass and surface waters dried up, Stokes cannot be taxed with failing to know that the promise was sometimes a grim and fatal joke, nor could he know that more than 125 years later his Plains of Promise would still be innocent of tapering spires of any sort.

Leichhardt, crossing the Gulf plains in the winter of 1845, was far more interested in new species of flora and fauna than in any potential the land held for pastoral purposes, but his party encountered no real difficulties. A. C. Gregory, travelling from the Victoria River districts of Western Australia in 1848, also crossed the Gulf country in the dry season, but being an experienced bushman with a good eye for country, he noted certain drawbacks which, he felt, offered 'no inducement to settlement'. His warnings that both widespread inundations and droughts probably prevailed from time to time, and that the grasses made inferior pasture, went unheeded in the optimism of the times.

## THE FIRST SETTLEMENT

During the search for the Burke and Wills party in 1861-62, three land parties, supported by a vessel, were actually in the field during the summer months, but it appears that the northern climate chose this particular summer to present a smiling face. Journals of these expeditions record surprisingly few wet days, and none of the parties experienced the slightest difficulty because of boggy conditions. Landsborough's accounts were particularly rosy, and his reputation as a practical grazier assured their wide publicity. While it is probably overstating the case to say that settlement of the Gulf country resulted from accounts based solely on this one abnormally 'dry' wet season, it seems highly likely that had these parties encountered a more typical season, with boggy ground and flooded streams to impede their progress, the course of settlement might have been somewhat different.

Considering the remoteness of this part of Queensland, settlement came with amazing rapidity. By 1869 most of the country north of the present railway line carried sheep or cattle, although some of it was surely being used without benefit of a lease, for legal occupancy of pastoral land in these parts of Queensland has always been on a leasehold basis. The seasons had, on the whole, been good. The dry seasons had been normally severe, but in those days natural surface water supplies were more plentiful and lasted longer into the dry season than now. The disadvantages of being far from sources of supply and markets were beginning to make themselves felt, and a severe financial recession in the State as a whole dried up the credit on which pastoral expansion had been based. These factors alone would probably have caused serious curtailment of operations in the North-west, but now the climate showed its full hand.

The summer season of 1869-70 may well have been the wettest the North-west has ever known; no records were kept. Edward Palmer (1903: 163), of Canobie Station on the middle Cloncurry, described conditions succinctly but graphically: '. . . no such flood was ever dreamt of, or has ever been seen since; it rained all January, February, and most of March, and the rivers covered all of the plain country.' This climatic calamity, coming on top of financial stringencies, caused virtual abandonment of the entire region; only a few stations managed to last through the bad times. The Gulf country holds the rare and certainly questionable distinction of being one of the few parts of Northern and Central

Australia to be abandoned because of too much water. By the time graziers reoccupied the abandoned country a few years later, they had acquired a wary respect for the country and its environment, although this new-found respect did not go as far as understanding. Now, however, they chose their land with an eye to permanent water, safety from floods, and proximity to such transportation as existed.

## THE CLIMATE

Few who live under it have kind words for the climate of Northern and Western Queensland. It has been described as 'hard on horses and men, hell on women and dogs'; it makes a mockery of the term 'normal season' and must be reckoned a sociological factor of no small importance. To the southern Australian and the immigrant, the climate is considered 'tropical', since much of Queensland lies north of that magical line $23\frac{1}{2}$°S.—a line which the Rockhampton Chamber of Commerce has recently shifted a mile or so north to make it coincide with a pleasant roadside park. However, only a small portion of North-west Queensland enjoys a truly tropical climate, and that only the tropical savannah variety in which most of the annual precipitation falls during a few summer months. The bulk of the North-west lives and works under semi-arid steppe and desert conditions.

Europeans have found two aspects of this climatic regime particularly frustrating: marked seasonality and variability of rainfall (Table 1). About three-fourths of the annual rainfall of 20 to 30 inches falls in one-third of the year, from December to March. This seasonality seems to decrease inland, so that three 'zones of seasonality' can be distinguished (Figure 13). In the districts adjacent to the Gulf of Carpentaria more than 80 per cent of the year's rain falls between December and March, while in the extreme south-western portion of the region not quite 70 per cent falls during these months.

The figures alone might suggest that the inland areas benefit from a somewhat more even distribution of rainfall throughout the year, but the annual averages at these stations are so low that the additional spread in terms of time is really a serious disadvantage. The Gulf country would seem to have the best of it, but again statistical appearances are misleading, for the 30 to 60 per cent higher rainfall in this district, coming as it does at the period of maximum temperatures, promotes such rapid growth of pasture grasses that they become rank, coarse, and of low nutritive value.

TABLE 1

*Some Characteristics of Rainfall in North-west Queensland*

Stations listed generally from North to South

| | Record | Annual | Dec.-Mar. | Above Average | | Below Average | | Average | Variability |
|---|---|---|---|---|---|---|---|---|---|
| | Years | Inch | % | Years | Extreme | Years | Extreme | Inch | % |
| Croydon | 49 | 28·35 | 84·4 | 24 | +23·78 | 25 | −14·96 | 6·10 | 21·5 |
| Normanton | 66 | 37·57 | 87·0 | 36 | +22·64 | 30 | −23·63 | 8·51 | 22·7 |
| Burketown | 65 | 27·61 | 87·3 | 30 | +48·38 | 35 | −20·23 | 10·36 | 37·5 |
| Lawn Hill | 41 | 20·11 | 81·9 | 20 | +13·85 | 21 | −12·35 | 7·16 | 35·6 |
| Donors Hill | 48 | 25·51 | 81·8 | 16 | +29·95 | 32 | −11·72 | 7·54 | 29·6 |
| Iffley | 43 | 21·78 | 75·0 | 22 | +22·43 | 21 | −14·06 | 6·49 | 29·8 |
| Millungera | 40 | 18·79 | 79·2 | 20 | +16·13 | 20 | −14·77 | 5·61 | 29·9 |
| Camooweal | 60 | 14·96 | 75·1 | 28 | +13·37 | 32 | −8·64 | 4·70 | 31·4 |
| Lake Nash | 40 | 12·27 | 73·9 | 14 | +18·64 | 26 | −9·25 | 5·17 | 42·1 |
| Cloncurry | 68 | 17·96 | 76·3 | 30 | +23·27 | 38 | −12·57 | 5·94 | 33·1 |
| Richmond | 48 | 17·80 | 73·1 | 21 | +15·59 | 27 | −12·98 | 6·86 | 38·5 |
| Dagworth | 54 | 15·95 | 70·5 | 22 | +21·21 | 32 | −12·03 | 6·17 | 38·7 |
| Carandotta | 56 | 9·82 | 64·2 | 26 | +16·05 | 30 | −8·64 | 4·17 | 42·4 |

In fact, it is the country in the middle 'zone of seasonality' which produces the best pasture, and portions of this area are far more prosperous than either of the others.

This seasonality difference has long since been translated into a land use pattern. In the 1860s the Gulf country was stocked with sheep, but they did poorly on the wetter, coarser Gulf pastures, and after the abandonment following the summer floods of 1869-70 restocking was with cattle. A further attempt to run sheep in this district in the 1920s and early 1930s ended in failure. On the other hand, the downs country of the upper Flinders and Cloncurry is the best sheep country in this part of Queensland. Farther westward, however, the ability of an area to carry sheep rather than cattle depends more on geology than climate: the rough, stony and dry ranges between Cloncurry and Mt Isa carry only cattle, but where black soil downs country exists in any quantity sheep can be successful, if water supplies permit. Carandotta, the driest of the stations listed, with only 10 inches per year, has long carried both sheep and cattle.

*Adaptations to Seasonality*

The markedly seasonal distribution of rainfall has had a profound effect on the manner in which the pastoral industry conducts its operations, and on the type of man who came to North-west Queensland. Generally the heaviest of the rains are over by the end of March or mid-April, but it may be another three or four weeks before the ground is dry enough to make travel possible and temperatures have moderated sufficiently to permit the working of stock. While rain may not preclude travel until November or December, by the end of October it is usually too hot to work stock safely. This means that most of the work, which in a less seasonal climate could be spread over much of a year, must be crammed into six, or at the most seven, months.

On a station, the dry 'winter' months are a period of intense activity. On cattle stations the stock must be mustered, calves branded, and the turn-off sent to market. Stock work on these large properties (2,000-5,000 square miles) is carried on from central camps which shift from one part of the property to another, and during the winter months employees concerned with handling stock rarely come to the head station. On stations running sheep, the dry season is the time shearing must be done, the wool sent to market and surplus sheep mustered and sent away. In all cases, major capital additions such as water improvements, fencing, construction of buildings and the like must be undertaken in the winter and completed before the wet season. Always supplies of all sorts, from aspirin to barbed wire, must be laid in against the coming wet season.

Summer is of necessity a slack period. Some maintenance work may be done at or close to the head station, such as repair of saddles, packs and other stock-working gear, repair of buildings and overhaul of mechanical equipment, but little can be done on the property as a whole. Station tracks and roads are often impassable for weeks, and communication is maintained by radio and aeroplane, weather permitting. Sometimes travel by vehicle is possible during the summer, always subject to uncertainty and prolonged interruption. The summer months are the traditional vacation period for managerial and other administrative staff, for not only are their duties at a minimum, but they are afforded an opportunity to miss at least some of the summer heat.

Organization of activities on such a seasonal basis makes unusual demands upon the labour force and to a very real extent determines the character of the labour force itself. Station managers and owners naturally wish to hire most of their labour for the winter months only. This policy is shared by contractors who make most of the important capital improvements, such as fencing, water supply, buildings and the like.

Seasonal labour, particularly in the pastoral industry, is an Australian tradition, but is neither efficient nor stable. Composed largely of men who are by nature drifters, this labour force is precisely that portion of the Australian population which would *not* be chosen to generate stability and solidarity. Fortunately for these men, the northern environment provides employment opportunities which match their inclinations so admirably. Nevertheless, the development of Northern Australia has been seriously hampered by a conspicuous lack of community feeling in its labour force, a labour force which is tailored to the environment.

*Consequences of Variability*

While the seasonal occurrence of the rainfall causes many problems for the North-western resident, the variability of the rainfall from year to year can spell success or disaster. If rain in almost any quantity is assured, careful planning can mitigate some of the disadvantages of unequal distribution; but if both the total amount received and the time of the year when it begins are variable, planning is much less likely to fit the situation as it develops. Thus in North-west Queensland, as in most of Northern and Central Australia, variability of rainfall is a most important environmental factor.

The rainfall pattern is variable almost to the point of paradox. Consider these variations over a four-year period at Burketown, where the fifty-year (1887-1937) average annual rainfall in 27·35 inches:

TABLE 2

*Burketown Rainfall Departure from Average*

| Year | Rainfall | Departure from Average | |
|---|---|---|---|
| (Jan-Dec) | inch | inch | % |
| 1891 | 66·14 | +38·79 | 139 |
| 1892 | 12·84 | −14·51 | 52 |
| 1893 | 11·31 | −16·04 | 58 |
| 1894 | 75·99 | +48·64 | 175 |

How can a grazier plan his work to allow for seasons of this sort? This is one of the environmental puzzles the people of the North-west have been trying to solve for over a century.

In no portion of the North-west is rainfall variability less than 20 per cent of the average (Table 2). Variability is least in the north-eastern portion of the region and increases south and westward. The greatest variability is associated with the driest climates; at Lake Nash and Carandotta, with averages of 12·27 inches and 9·82 inches respectively, the variability is slightly over 40 per cent.

At some stations in the region, variations above the average yield rather high annual totals for Australia, although no year has received more than 190 per cent of the average. The wettest years are the result of an occasional tropical cyclone or of a particularly well-developed monsoonal flow, controlled by the pressure systems of Asia. Both conditions are the exception rather than the rule, and their infrequency in recent years has led the residents to speculate that the climate is becoming drier. Moderate amounts of rain above the average generally assure a good season, with grass and water in abundance. Marked excess, however, brings no bonus, for then the land will be boggy for prolonged periods, flood losses in terms of stock, fencing, roads and other improvements will be high, and pastures will be coarse and low in nutritive value.

Variations below the average are, on the other hand, a much more serious matter, particularly in areas where the annual average is under 20 inches. Such failure of the rains upon which both pasture and water supply depend so heavily can mean disaster, for the minimum amount needed, rather than the maximum permissible, is the real limiting factor. Over most of the North-west, as in dry climates almost everywhere, more than half the year's rainfall is below average.

A further and more serious aspect of this variability is the lamentable tendency for one poor season to follow another. Thus the rainfall registered at Lake Nash fell below the average during fourteen of the sixteen years between 1922 and 1937. Under such conditions pasture disappears, all soil moisture is exhausted, surface water supplies give out and even the ground water supply may be affected. Then significantly larger falls are required to start the pastures, and it may take more than a decade for pastures and water supplies to recover fully from a prolonged drought of this sort.

## CLOSER SETTLEMENT POLICY

Seasonality and variability of rainfall mean little to the person who merely reads of them, but to the people who make their homes and livelihood under these conditions they present the ultimate challenge. Unfortunately these aspects of the physical environment have received insufficient attention in the evolution of the long-term lease from the Crown, with whom title rests. Queensland governmental policy has always held that reasonably cheap land should be available to all qualified persons wishing it, and the 'closer settlement' process is the machinery by which this is accomplished.

Pastoral leases provide for 'resumption' of certain percentages of the land after the lease has been in effect a specified number of years, providing there is a real demand by intending settlers. These resumed lands are surveyed into properties of suitable size and distributed by ballot to qualified persons wishing to take up land. It means, of course, gradual diminution of the larger properties, and if properly prosecuted this is often desirable. However, it does not always so operate, as the following example shows.

Despite the early failure of sheep on the country immediately adjacent to the Gulf of Carpentaria, pressure for closer settlement during and after World War I led the State government of the day to resume substantial areas from a number of properties and re-lease them in small blocks (20,000 to 60,000 acres) as sheep properties. One of these was a property near Burketown of about

30,000 acres, owned by one Reg Murray, who by the early 1930s was running about 12,000 sheep with rather mixed success. While the Gulf country proudly boasts that it never knows a 'king' drought, the wet season of 1934-35 broke very late; *no* rain was recorded at Burketown for December 1934, the only year on record to show such a deficit. Conditions did in fact approach a drought and Mr Murray, along with other graziers in the district, was forced to haul water from whatever sources were available to keep his sheep alive. Even so, some losses were suffered among old sheep and lambs.

Late in January the season broke with a vengeance, and more than 10 inches of rain were recorded in a single 24-hour period at a number of stations in the Burketown district; the January (1935) total at Burketown was 15·37 inches, more than 7 inches over the average for that month. This deluge was preceded by several days of strong northerly winds, which tended to pile water into the shallow southern end of the Gulf of Carpentaria, and coincided with a period of high tides. As a result the abnormally heavy run-off could not get away rapidly, and vast areas of the even coastal plain were deeply flooded; Mr Murray reported water 11 feet deep at his homestead.

Cattle seem to exhibit a certain instinct for self-preservation when confronted by rising flood waters, and usually seek higher ground ahead of the spreading water. Sheep, however, seem to be imbued with no such common sense: they stand dumbly in small groups and watch the water rise around them until they are finally swept away, or dash about aimlessly until their wool becomes saturated and they are bogged down in mud and water. So it was with Mr Murray's 12,000 sheep. He estimates that about 5,000 were lost during the first forty-eight hours of the flood, drowned and swept away by the water. Over the next month many others succumbed to bogging and blowflies. When he was finally able to muster his water-logged paddocks, Mr Murray could find less than 4,000 sheep. For all practical purposes he was out of the sheep business.

## CONCLUSION

Such an environment profoundly affects the lives of all who experience it. Their births may be planned to take advantage of the cooler winter, their burials delayed because their graves fill with water. It is not strange that these people regard the environment, which they reduce to mean simply climate, as all powerful:

it appears to bring or deny the wherewithal for happiness. Nor is it strange that some of this rubs off on those whose business it is to investigate these lands and people.

The fact is, of course, that in these parts of Australia the environment strictly limits human activities, and the activities in which Europeans have chosen to engage do not fit the environment over-well. This, of course, cannot be blamed on the environment, but rather on the cultural background of those settling there, a background which included small knowledge of how to deal with such a pronounced seasonal environment.

The white man has posted some notable successes, to be sure, but these have been based largely on individual persistence, hard work, luck, and intestinal fortitude. Fortunately the end is not yet. Considerable bodies of knowledge and technology relative to this and similar parts of the world are rapidly accumulating. Their sensible application by those dealing with Northern and Central Australia—government bureaucrat and scientist, pastoral and mining company, bank official and grazier—may yet induce the North to yield the largesse it has so long promised but yet withholds.

## BIBLIOGRAPHY

CARSTENSZOON'S *Journal*, IN HEERES, J. E. (1899). *The Part Borne by the Dutch in the Discovery of Australia, 1606-1765*. Luzack and Co., London.

GREGORY, A. C. (1858). 'Journal of the North Australian Exploring Expedition', *Jour. Royal Geog. Soc. London*, **28**: 1-137.

PALMER, EDWARD (1903). *Early days in North Queensland*. Angus and Robertson, Sydney.

PRICE, A. G. (1939). *White Settlers in the Tropics*. American Geographical Society, Special Publication, No. 23, New York.

STOKES, J. L. (1846). *Discoveries in Australia*. T. and W. Boone, London. 2 vols.

FAY GALE

# A Changing Aboriginal Population

*'The number of mixed bloods is everywhere on the advance. . . .'*
A. GRENFELL PRICE *White Settlers and Native Peoples*

ON THE SHORES of Lake Alexandrina, at the mouth of the River Murray in South Australia, there is an Aboriginal reserve, known as Point McLeay. The area has been a very important camping ground for Aborigines since 'the dreaming'. With the arrival of Europeans, radical changes took place amongst the Aborigines who came to this sacred site. The full-bloods declined in numbers, but the mixed-bloods who appeared eventually increased more rapidly than the full-bloods decreased.

The changing fortunes of the Point McLeay Aboriginal population are analysed here in the hope that they may give some indication of the demographic changes which have taken place generally amongst Australia's indigenous minority since Europeans first settled in this country.

Unfortunately, such a study cannot be made from census data. Aborigines were excluded by the provisions of the Commonwealth Constitution (at federation in 1901) from the population to be counted by census. For this reason, they have not been included officially in any Commonwealth census, although Aborigines designated as 'half-caste or less' have been included. Often in the past and certainly in the last two censuses, full-blood Aborigines have been counted in most areas because it is not possible to distinguish them from half-castes. But this approach has produced, for Australia's Aborigines, population figures which are rather inadequate and poorly classified. The results have not been satisfactory for the purpose of making a serious study of population trends over a period of time.

Complete census data on Aborigines will be available from the 1971 census onwards, because the Commonwealth Constitution was changed by referendum in 1967, so that all Aborigines could be included officially in an Australian census. But there can never be reliable census data to make possible a study of the Aboriginal population in the past. Such a study can be made only by the reconstruction of a population, through original field research. Such a method of obtaining data presents the almost insurmountable problem of establishing a total or 'closed population'. But, because there is no other way of studying this past population, an attempt has been made to acquire demographic data for one group so that some assessment can be made of the changes in the Aboriginal population since the arrival of the white man. This study does not guarantee to have established a 'closed population' but it has come as near to this as may ever be possible.

After exploratory work in several areas, the Point McLeay reserve was chosen for this sample demographic study. There were two main reasons for choosing Point McLeay. The Aborigines in this area have had a long history of European contact. The first recorded white visitors were from a shipwreck in 1838, but there is evidence that the sealers and whalers, based on Kangaroo Island from 1803 to 1836, took Aboriginal women from the Aboriginal ancestors of the Point McLeay people. The first permanent white settlement came in 1859, when a mission was established to 'protect' the Aborigines from European contact.

The second reason for selecting Point McLeay for this study was that the records kept for this group of Aborigines seemed more detailed and reliable than those of any other group in Australia. One of the most useful records is a Register of Births, Deaths and Marriages which has been kept, more or less continuously, since 1859. It would have been relatively simple to make a study of population changes, based solely on the entries in this Register. But this would not have allowed for the factor of migration away from the reserve. Far more extensive work than that was necessary to establish a reasonably reliable population group.

For the purposes of this study, the compilation of a genealogy was the first step towards the establishment of the 'closed' Aboriginal population, of Point McLeay origin. This was done by interviewing Aborigines who recognize themselves as descendants of the original inhabitants of the Point McLeay mission, and are happy to acknowledge that they or their forebears came from 'the Mission'. This was not an easy task, for today they are a very scattered people. Field work had to be done in many parts of the State to establish the genealogy.

Each individual was recorded on a numbered, cross-referenced card. This method of recording was adopted to allow for the numerous changes and additions which are involved in the genealogical recording of a dispersed group. It is unlikely that the data could have been obtained or satisfactorily analysed had the more common anthropological method of recording genealogies been used in this study.

Though geographically scattered, the Point McLeay people are a self-defined population. They recognize each other as members of a group having this common origin. Even in Adelaide, they have more social dealings with each other than with any other Aborigines or white Australians. They marry within their own group more frequently than do any other Aboriginal groups in South Australia. 'Gossip channels' keep each member of the Point McLeay group aware of the activities of the other members of the group. It was not, therefore, as difficult as one might expect to reconstruct a genealogical study from such a geographically dispersed people because they are a group whose members are both genetically and functionally connected. Nevertheless, the collection of this data took several years.

No matter how satisfactory a genealogy may appear when constructed from field informants, it is quite inadequate and unreliable for use in demographic analysis. Memories cannot be guaranteed. Before any study could be made of the genealogies, much laborious checking was necessary. Interestingly, when Aborigines knew that their information was being checked against that of other informants and official sources, they were much more precise and less inclined to romanticize their past.

Apart from the Register of Births, Deaths and Marriages, already referred to, many checks were used. An extensive and well-documented record of these people had been made by N. B. Tindale in the 1930s. This material has not been published, but the original field notes are kept in the South Australian Museum. R. and C. Berndt also collected genealogical information about these people, but this material is not published and was not available to the writer. The first missionary, George Taplin, stationed at Point McLeay, kept records which are now preserved in the South Australian Archives. The Department of Aboriginal Affairs continued the precedent of population recording when it took over the mission in 1914. Further checks were made by obtaining access to the personal files and family case histories held at the Department of Aboriginal Affairs, the Department of Social Welfare, the Gaols and Prisons Department and the three major government hospitals in Adelaide. There were very few Point McLeay people who were not recorded in at least one of these places.

Thus, it was possible to check the information obtained from Aborigines against official and semi-official records. While the genealogical technique was the basis of this study, it can be seen that many controls were used to check faulty or romanticized memories.

Almost a century is covered by this study: from 1 January 1870 to 31 December 1964. There are several reasons why 1870 was chosen as the base line. Although entries in the Point McLeay Register commenced in 1859, it was not until 1870 that age-at-death was entered in all or nearly all of the cases. In the absence of earlier birth registrations, age-at-death was necessary to aid in the calculation of the total population for any one year. Furthermore, there were a few older Aborigines in the age-group eighty to ninety-five years, who were still alive when the field study was made. They helped to establish approximate birth dates for checking with the Register. Although they could not remember all individuals or actual dates, when prompted with other information, they could tell whether a certain person was younger or older than another individual. By this means, it was possible to assign births, with reasonable accuracy, not to an exact year but to a five-year period. Even so, one cannot depend too much on the accuracy of the population data acquired for the period from 1870 to 1909. Nevertheless, the older informants were very valuable and, because of this and the other sources of information, it seemed reasonable to take the study back to 1870.

The completed genealogy eventually included 2,625 Aborigines of Point McLeay origin. Only 1,388 of these were entered in the Register of Births, Deaths and Marriages for the reserve. This difference of 1,237 individuals shows the significance of the migration factor. Although they were not on the Register, nearly all of the ancestors of the present study were recorded in the Tindale collection.

The following demographic study does not, unfortunately, include all of the 2,625 individuals traced. After extensive checking it was decided that 289 Aborigines must be omitted from the analysis. This decision was made because the information obtained for these 289 individuals was either insufficient or could not be verified from other sources. A few had married Aborigines from other States and had moved away from South Australia and their relatives knew little about them. They did not appear on any official records in any of the State departments and therefore information given by relatives could not be checked. Approximately one-third of them had moved away from the reserve into the normal community and ceased to have contact with any of the remainder of the group. Some two-thirds had become assimilated into the general community through marriage to white spouses. The actual figures suggest that the rate of assimilation or 'passing over' as the Aborigines call it, has been in the vicinity of 10 per cent over almost a century.

Thus 289 individuals have been excluded and this study is limited to 2,336 self-defined Aborigines from Point McLeay, for whom reasonably reliable data have been collected. It is possible to estimate the effect of the exclusion of this 10 per cent on the following analysis. Since nearly all of those excluded have been assimilated into the white community, it seems reasonable to suggest that their family size must have been nearer to the white norm than to the remainder of the Aboriginal community discussed in this paper.

## BIRTH- AND DEATH-RATES, 1870-1964

In 1870 the population of the Point McLeay group of Aborigines numbered 271 individuals. Even at that time they did not all live on the reserve, but nearly all were recorded in the Register because entries were made originally not only for the resident Aborigines but also for related Aborigines who moved backwards and forwards between the reserve and neighbouring areas. The figure of 271 was obtained by analysing age-at-death, as nearly as possible, for each person who died after 1870.

A study of Table 1 and its graphic representation in Figure 14 will reveal the changes in the Aboriginal population of this region over almost a century of time.

It is evident that from 1870 until 1894 the overall population of Point McLeay Aborigines was declining. The death-rate was higher than the birth-rate throughout this period, and probably in the preceding period also. It is not possible to push back the graph any earlier than 1870, but archival material suggests that the population in 1859, when the mission commenced, was much higher than in 1870. It seems evident that the Aboriginal population declined from the time of first white contact right to the end of the nineteenth century. Historic records from many parts of Australia support this pattern of a rapidly declining population resulting from a high death-rate, for something like fifty years after white settlement. By that time, either the Aborigines died

TABLE 1

*Population of Point McLeay, 1870–1964*

| Year | Births | Deaths | Total Population | Birth-rate | Death-rate |
|---|---|---|---|---|---|
| 1870 | | | 271 | | |
| 1870–74 | 25 | 36 | 260 | 19 | 28 |
| 1875–79 | 50 | 66 | 244 | 41 | 54 |
| 1880–84 | 47 | 50 | 241 | 39 | 41 |
| 1885–89 | 32 | 36 | 237 | 27 | 30 |
| 1890–94 | 64 | 59 | 242 | 53 | 49 |
| 1895–99 | 78 | 58 | 262 | 60 | 44 |
| 1900–04 | 64 | 45 | 281 | 46 | 32 |
| 1905–09 | 85 | 87 | 279 | 61 | 62 |
| 1910–14 | 83 | 60 | 302 | 55 | 40 |
| 1915–19 | 76 | 48 | 330 | 46 | 29 |
| 1920–24 | 84 | 49 | 365 | 46 | 27 |
| 1925–29 | 96 | 44 | 417 | 46 | 21 |
| 1930–34 | 130 | 36 | 511 | 51 | 14 |
| 1935–39 | 164 | 69 | 606 | 54 | 23 |
| 1940–44 | 169 | 65 | 710 | 48 | 18 |
| 1945–49 | 166 | 44 | 832 | 40 | 11 |
| 1950–54 | 230 | 64 | 998 | 46 | 13 |
| 1955–59 | 234 | 53 | 1,179 | 40 | 9 |
| 1960–64 | 188 | 39 | 1,328 | 28 | 6 |

out or some form of population adjustment took place and gradually the population began to increase again. For Point McLeay people, and indeed for most Aboriginal groups in Southern Australia, the adjustment meant the dying out of the full-blood but the increase of the mixed-blood numbers. All but three of Point McLeay's living 1,328 Aborigines have some white ancestry.

In parts of Central and Northern Australia where contact with whites has been much slower and more spasmodic, even the full-bloods have had time to adjust to introduced diseases and foods and their population numbers are now increasing in some areas. But not so in the south. The pattern of change, exhibited by the Point McLeay group, is probably not very different from that shown by other groups. It could probably be applied to most of Southern and Eastern Australia.

Figure 14 shows an erratic graph for the Aboriginal population in the early period from 1870 until 1910. There are two reasons for this erraticism. Total numbers are quite small, in the earlier decades, so that minor variations show up significantly, whereas in

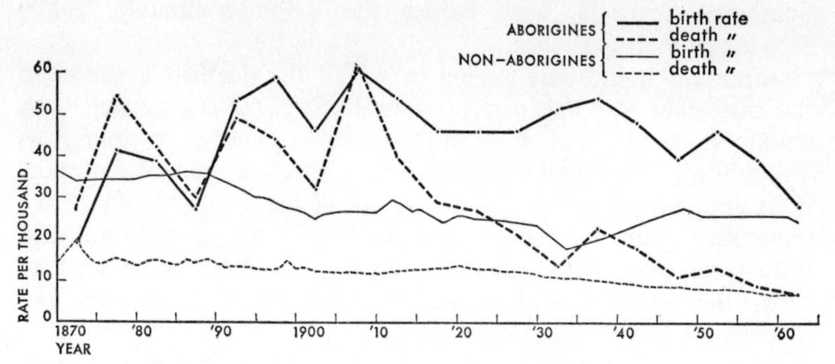


FIG. 14 A comparison of Aboriginal and white population change, Australia, 1870-1964

the later decades, where numbers are greater, only very definite variations are able to change the shape of the graph. Furthermore, the data for the earlier period is less reliable than that obtained for the later decades, so that fluctuations are more likely in earlier results. Nevertheless, the general pattern, irrespective of the exact figures, is no doubt correct. Death-rates certainly surpassed birth-rates for most of last century.

The parallel peaking of birth- and death-rates is to be expected because of the high rate of infant mortality. It appears that, after 1895, a steadily increasing birth-rate overtook the high death-rate and led to a natural increase in the population. This upward change in the graph occurs earlier than anticipated. Even as late as the 1930s it was generally agreed that Aborigines were dying out. Several articles and newspaper reports of the period discussed Aborigines as 'a dying race'. In 1938 Daisy Bates published a book entitled *The Passing of the Aborigines*. Indeed, one of the reasons that Aborigines were not included in the Commonwealth Constitution in 1901 was the firm consensus of opinion that Aborigines were rapidly dying out. Certainly the full-bloods, as a distinctive race, were diminishing in eastern and southern Australia. Increasing interbreeding with whites ensured that they did not remain a separate genetic group. It was true, therefore, that the full-blood Aborigines were dying out in the areas of rapid European settlement. But the mixed-bloods were never socially separated from the full-bloods. The mixed-bloods continued to identify themselves as Aborigines, and they certainly were not dying out. By the turn of the century this new breed of Aborigines was on the increase in

Southern Australia, long before the white community really recognized their existence.

After 1895, the only period in which the death-rate exceeded the birth-rate was between 1905 and 1909. In this period, both whooping cough and respiratory infections, listed variously as 'bronchitis', 'pneumonia' or 'influenza', reached epidemic proportions and accounted for more than half of the deaths. Many of the remaining adult full-bloods and young full-blood children died during these epidemics. But the hybrids seemed to be able to resist the introduced diseases. The full-blood Aborigines had reached the last stages of their decline.

There was a considerable increase in the numbers of mixed-bloods from 1910 onwards. The birth-rate remained at the high level reached in the 1890s and, coupled with the decline in the death-rate, led to a gradual increase in the total population. The birth-rate remained high until after the Second World War. From 1910 to 1940 there was an acceleration in the rate of population increase. Since 1940 the birth-rate has gradually declined, but the high level of population increase has continued, because the death-rate has been declining much faster than the birth-rate.

The graph shown in Figure 14 does much to explain why there has been a constant migration away from the old Point McLeay reserve ever since *circa* 1920. There were simply too many people for such a small area (5,729 acres) of pastoral country. The sheer pressure of numbers was forcing their assimilation into the general community without any impetus from the actions of governmental authorities.

A comparison of the Point McLeay graph with the one for Australia as a whole, over the same period of time, shows that both the birth- and death-rates for the Aboriginal group have been considerably higher than the rates for the general population of Australia. It is probable that in most of the older, settled areas of Australia, the Aboriginal population has been increasing at a rate considerably greater than that of the white Australians for most of this century.

In contrast with the general Australian population, the birth-rate of this Aboriginal group—and, one expects, of most Aboriginal communities—remained high during the depression of the 1930s. For Aborigines living on missions and government reserves, economic pressures upon the family were no greater at this time than at any other time: free housing and rations were just as readily available then as before. Reserve dwellers did not suffer from loss of employment because so few of them had been employed in any case.

The Second World War set Aborigines on a very different course. Some Aborigines enlisted. Others were taken to employment on farms or in factories away from the reserves, and were often separated from their families. The birth-rate declined during these war years. For the first time, Aboriginal population trends began to reflect the same influences which were affecting the white community. Similarly, in the post-war period there was a marked increase in the birth-rate of each population. But, overall, the general trend is one of a decreasing Aboriginal birth-rate.

It is not possible to compare the figures for the Point McLeay Aboriginal group with those of the total Australian Aboriginal population because of the lack of sufficient census data. But it is possible to make comparisons with the New Zealand Maori population. Indeed, there is a much closer parallel between Point McLeay population figures and Maori statistics, for both birth- and death-rates, than exists between Point McLeay Aboriginal figures and general Australian population statistics.

Table 2 shows a comparison of birth- and death-rates for Maoris, taken from the New Zealand Official *Year Book*, and Point McLeay Aborigines during the three five-year periods from 1950 to 1964.

TABLE 2

*Maori and Aboriginal Birth- and Death-rates*

| | Birth-rates | | Death-rates | |
|---|---|---|---|---|
| Year | Maori | Pt McLeay Aborigines | Maori | Pt McLeay Aborigines |
| 1950–54 | 45 | 46 | 11 | 13 |
| 1955–59 | 45 | 40 | 9 | 9 |
| 1960–64 | 45 | 28 | 7 | 6 |

For the five-year period 1950-54 there is a close similarity between the Maori population figures and those collected for the Aboriginal group. In the present decade, however, the Aboriginal group has shown a considerable decrease in its birth-rate. The declining death-rate, resulting from improved medical services, is evident in both groups. But the Maori birth-rate is apparently remaining at a high level.

The declining fertility of the Aborigines from Point McLeay is a striking feature of these statistics and requires explanation. One might say that in these recent years, Aborigines have spread out in so many different directions, and have become so completely absorbed into the general community, that several may have been forgotten by their relatives and therefore not included in this genealogically-centred method of establishing a population. This does not seem very likely. Aborigines, especially the women, are extremely interested in their kin ties. Indeed, relatives and their activities provide one of the chief topics of conversation. Members, who acknowledge their origins, talk about those who have 'passed over' into the white community, and do their best to keep others informed about such recalcitrant relatives. It has already been explained that 10 per cent of the total number of individuals listed have been deliberately excluded from this analysis, because, although known to their relatives, they have become so identified with the white community that it was not possible to check information about them. It is doubtful whether the absorption has been higher than this figure of 10 per cent suggests.

But even if the 'loss' had been much higher than the 10 per cent already estimated, these assimilated Aborigines could not have become completely absorbed into the general community unless their patterns of behaviour (especially the size of their families) were more or less in line with those of the Europeans in the rest of the community. Thus, even their inclusion in the population statistics is unlikely to have altered the result to any great extent. Therefore, their exclusion, whether they amount to 10 per cent of the total population or more, is probably not greatly significant.

Undoubtedly, declining fertility in the present decade is due to increasing urbanization. Evidence for this can be seen in Adelaide which is the main urban centre. Whereas only 28 Aborigines of Point McLeay origin had come to live in Adelaide by 1956, 639 had moved into this city by 1966. Of these 639 persons, 365 had actually been born at Point McLeay. The remaining 274 were born of parents who had come from Point McLeay.

The declining fertility of the last decade has been due to the migration away from the reserve and identification of Aborigines with the general white population. The average family size of Aborigines who have remained on the reserve is considerably higher than the average family size of Aborigines who have moved to Adelaide. When Aborigines move from any reserve, or from a separate community on a town fringe into the 'outside world', they are faced with strong economic and social pressures to limit their families. One Aboriginal woman who had moved from Point McLeay to Adelaide explained this by saying: 'the mission is a breeding-up place. You can't go on having kids every year when you live in the city.'

Increasingly, Aborigines in the city are making conscious attempts to limit their family size. Certainly medical science has provided simpler and more adequate means of birth control in the last decade. These benefits are more readily available in the city than they are on the reserve. Aboriginal women in the city are becoming more aware of the practicability of family planning. The comparison of contraceptive methods provides a common topic of conversation amongst city Aborigines. These were never discussed, at least not freely, on the reserve. Since these data were obtained by a female research worker, it was possible to study and compare different attitudes on this subject. It was found that single girls and *de facto* wives in the city are just as conscious of the necessity to restrict births as are married women.

Intermarriages between Aborigines and Europeans are closely associated with the urban movement and have contributed to the decline in fertility. In the following table, the Aboriginal women who have borne children are grouped into fifteen-year cohorts on the basis of their year of birth. A comparison of their mates shows an increasing tendency towards interbreeding and intermarriage between Aboriginal women and European men.

TABLE 3

*Comparative Matings of Women who have Borne Children*

| Cohorts of Fertile Women | Origin of Mate (%) | | | |
|---|---|---|---|---|
| Birth Dates | Aboriginal | | European | |
| | Marriages | Liaisons | Marriages | Liaisons |
| 1885–99 | 72 | 7 | 8 | 13 |
| 1900–14 | 75 | 6 | 6 | 13 |
| 1915–29 | 62 | 11 | 9 | 18 |
| 1930–44 | 53 | 9 | 14 | 24 |
| 1945–59 | 37 | 13 | 39 | 13 |

Women in the last cohort are too young to have borne many children; only a few of the women have reached marriageable age.

Thus, the figures of the 1945-59 cohort are not complete. But the other cohorts show the trend to a decreasing percentage of Aboriginal marriages and increasing percentages of marriages and liaisons with whites.

It is obvious that interbreeding is not a recent phenomenon. Sexual relationships between Aboriginal women and white men have been common since the first Europeans arrived. Indeed, for the Point McLeay group at least, even marriage between Aborigines and Europeans has not been uncommon. But the numbers of marriages between Aborigines and Europeans is increasing. A significant factor in declining fertility may be the increasing intermarriage with whites. If the 1945-59 cohort continues to show the same mating pattern throughout the child-bearing period, then one can expect that intermarriage will increase and fertility will decline even further in the future.

Aboriginal women married to white men say that they must limit their family size because neither their neighbours nor their husband's relatives will accept them if they have too many children.

Intermarriage is not the only factor limiting fertility in the city in the present period. The social and economic elements in urbanization exert an independent influence on fertility rates, irrespective of whether marriages are mixed or not. In most families visited by the writer, the Aboriginal mother usually looked after the budget, whether her husband was white or Aboriginal. Typically, she often discussed the problem of making her income meet the ever rising costs of living. Family size naturally became a part of such increasing economic awareness. But neither budgeting nor family planning were common topics for conversation on the reserve.

Urbanization not only limits the family size of married women, but it also tends to lessen the number of children born to single Aboriginal girls. This does not necessarily mean that city living improves morality. It may be just that in the city there are more social pressures acting upon single girls who have children. Besides, they find it hard to support themselves financially in the city. This does not apply on the reserve where pregnancies in single girls tend to be treated as a social norm, and they are always supported economically. But in Adelaide, single girls seem more concerned either to induce miscarriages or to use some method of contraception. These practices are discussed freely and information is passed on amongst Aboriginal women in the city. Not so on the reserve.

Thus, one could say that urbanization, both through increasing intermarriage with Europeans and through the independent economic and social pressures of urban life, is definitely influencing Aboriginal fertility.

The increasing migration of Aborigines from reserves and country areas to the city must have a strong influence on Aboriginal population growth in the future. The urban effect may be more evident in the Point McLeay group than in many others because they have emigrated from their reserve at such an accelerated rate in recent years. By 1964, only 11 per cent (147 out of 1,328) of the Point McLeay group were still living on the reserve. Similarly, Aborigines from other reserves are moving to the city and declining fertility must take place as they become urbanized.

A study of the population of Point McLeay has thus shown three distinct phases. A population which was probably static in pre-European times, began to decline rapidly after its first contact with Europeans. Indeed, so rapid was the decline that virtual extermination seemed inevitable. Then gradually a change appeared and the population began to revive until eventually its rate of growth became much greater than that of the rest of the white population. In the final phases, it appears that the Aboriginal population is gradually settling down again to exhibit much the same trends as those shown by the rest of the Australian community.

A similar migration movement and increasing urbanization are evident amongst the Maoris (Metge 1964), but they have not yet reached the high percentage rate of the Point McLeay Aboriginal migration. As yet, Maori urbanization does not appear to have had much effect on the high rate of Maori fertility; this is still running at a rate equivalent to that of Point McLeay a decade ago.

## AGE AND SEX COMPOSITION, 1930-1946

The pyramids in Figure 15 show the age and sex composition of the Point McLeay Aborigines in five-year groups over a span of thirty-five years. These are compared with the total Australian population at the 1961 census. The Aboriginal pyramids show a very youthful age structure, but this is to be expected in view of the high birth-rate. But the age structure has gradually changed since 1930. It can be seen that, in the period 1935-39, some 51 per cent of the Aboriginal population was under the age of fifteen years, whereas, twenty-five years later, in the period 1960-64,

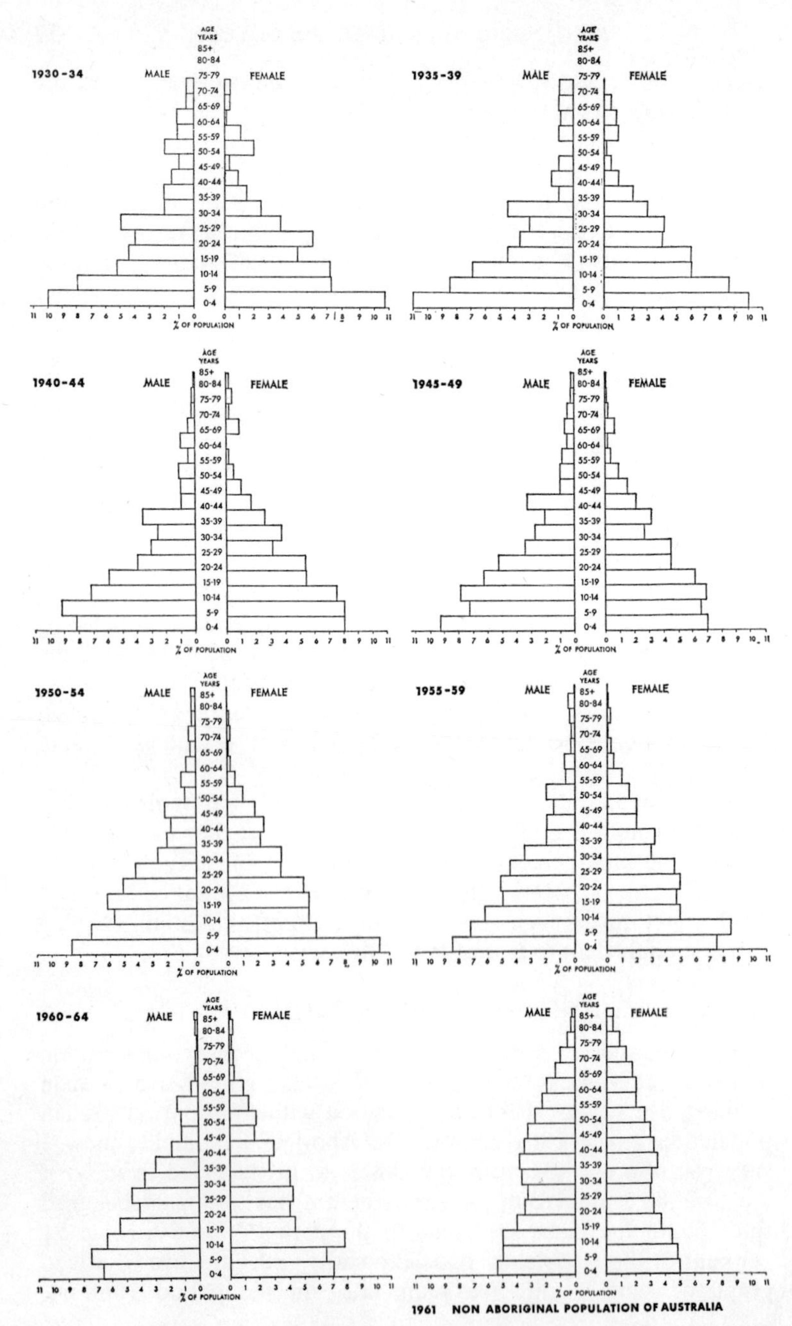


FIG. 15 Changing age structure of Aborigines between 1930 and 1964, compared with the age structure of other Australians, 1961

only 42 per cent of the Aboriginal population was under fifteen years of age. A change is evident, but there is still a vast difference between the age structure of the Aboriginal population and that of the general Australian population. According to the 1961 census only 30 per cent of the total Australian population was under fifteen years of age.

Conversely, there are proportionately fewer old people in the Aboriginal population. In 1960-64 only 2 per cent of the Aboriginal population was over sixty-five years of age, whereas in 1961 8·5 per cent of the Australian general population was over sixty-five years of age. This means more child endowment and fewer age pensions for Aborigines, but the social consequences of such an age structure are more significant than just the matter of pensions.

The Aboriginal masculinity ratio is slightly different from that of the Australian norm. The ratio of males to females in the older Aboriginal age groups is significantly different from the general population. In the Aboriginal group males predominate at the older ages. The reverse is the case in the Australian population as a whole. The Point McLeay group is showing a change in this predominance of older males. Whereas in 1930 the male numbers were greater than female numbers at the age of thirty-five years, by 1950 male predominance was not evident until the age of fifty-five years. In the last five-year period, 1960-64, the masculinity ratio was no longer any higher in the older age groups. This trend is no doubt due to historic factors. A study of the Death Register and personal case histories has shown that there were more female deaths due to child-birth in the early part of this century than there have been in recent years. The data obtained for this study have also shown that the female infant mortality rate at the end of last century was higher than the male rate. The reasons for this are not clear. It is possible that infanticide was practised more on female babies than on male.

There was evidently a high level of masculinity in the past. This has steadily dwindled in recent years, so that, by 1964, the male-female ratio was approximately equal in the Aboriginal population. One wonders whether the masculinity ratio was significantly higher in the Aboriginal population before Europeans came to Australia. It appears that female deaths due to child-birth were high after the Europeans arrived. But were they just as high before interbreeding with whites occurred? Masculinity in the older age-groups may have been a part of traditional Aboriginal population structure, or it may have been a temporary result of initial contact with Europeans. Certainly high masculinity does seem to be related to declining populations. When the Aboriginal was dying out, the masculinity was highest. As the population revived, the masculinity declined. Perhaps the high masculinity was merely a temporary feature of culture change which, along with other features of the population, has taken just over a century to right itself.

## FERTILITY

The high rate of natural increase amongst the Point McLeay Aborigines suggests a high level of fertility. Crude fertility rates were worked out on the basis of births per thousand individuals in the population. The analysis worked out on this basis produced the general trends shown in Figure 14. However, the Point McLeay data are sufficiently detailed to allow a more precise measurement of fertility.

It is possible in this paper to study the actual fertility experience of women over their entire reproductive period. The data contain both numbers of children born to each woman and her age at the birth of each child. Thus a table can be drawn up showing the average number of children per woman at any particular age.

The cohort approach is used so that changes in fertility over a period of time can be observed. All women in the population have been grouped into fifteen-year cohorts on the basis of their dates of birth. The collected statistics make it possible to go back as far as 1870. Thus, the first cohort consists of women who were born between 1870 and 1884. The second cohort consists of women who were born between 1885 and 1889, and so on. The last cohort takes women born between 1930 and 1944. It is of little value to include the women born in the next period, between 1945 and 1959, since they had scarcely commenced their reproductive cycle by the time this study was made, and certainly they had not borne sufficient children to show any particular trends.

Table 4 shows the average number of children born to these cohorts of women by certain ages. These statistics are worked out on the basis of the average number of children born in each five-year period to the number of women alive in that five-year age group. The number of women who died have been deducted from the relevant five-year age group. Female deaths during the reproductive years were quite significant during the latter part of last century and early in this century. In the cohort of women born

TABLE 4

*The Average Number of Children per Woman*

| Age group | Birth Cohorts of Women | | | | |
|---|---|---|---|---|---|
| | 1870–84 | 1885–99 | 1900–14 | 1915–29 | 1930–44 |
| 10–14 | 0·03 | 0·07 | 0·07 | 0·03 | 0·09 |
| 15–19 | 0·75 | 0·65 | 0·60 | 0·34 | 0·74 |
| 20–24 | 2·06 | 1·74 | 1·62 | 1·59 | 1·99 |
| 25–29 | 3·15 | 2·85 | 3·57 | 3·14 | 3·44 |
| 30–34 | 4·34 | 3·69 | 5·40 | 4·76 | 4·86 |
| 35–39 | 5·14 | 4·46 | 6·62 | 5·86 | — |
| 40–44 | 5·75 | 4·73 | 7·47 | 6·13 | — |
| 45–49 | 5·84 | 4·82 | 7·52 | 6·13 | — |

between 1870 and 1884 there were thirty-six women who reached the age of fourteen years, but only twenty-one who survived to the age of forty-five years. Thus, the number of children born to women in each five-year age group is averaged against the number of women who actually survived in each five-year age group.

Table 4 shows the actual reproductive history of women born since 1870. Women in each cohort bear most of their children between the ages of twenty and thirty-four years, but some women have children throughout their menstrual life. A few women had borne children before they were fifteen years of age and a few gave birth after they had turned forty-five years of age. The figures suggest that Aboriginal women have always commenced their families at a fairly early age and have continued to have children through most of their fertile period. Early marriage for girls was recorded in the literature of last century. Pregnancy before marriage has not been infrequent. These factors are related to the fairly high level of fertility evident in each cohort and the period of time over which women have their children. It is not clear whether the high levels of fertility and the long reproductive period were aspects of Aboriginal behaviour in pre-European Australia or whether they resulted from culture change and inter-breeding with whites.

The cohort approach used in Table 4 makes it possible to identify the actual periods of decline and increase in Aboriginal fertility.

It seems apparent that fertility was declining until well into the twentieth century. The cohort of women born between 1885 and

1899 produced the lowest average number of children per woman. It is not possible to tell when the decline began, but it certainly reached its lowest level with this particular group. Even at this relatively low level, fertility was still high in comparison with the general Australian population. Thus, the declining total population in the later part of last century was due more to increasing mortality than to declining fertility. By contrast, the high rate of population growth after 1910 was due to the combined factors of decreasing mortality and increasing fertility. The cohort of women born between 1900 and 1914 produced the greatest average numbers of children per woman. Though the reproductive history for later age groups is not complete, it appears that women born after 1915 tended to have smaller families than those born between 1900 and 1914.

Thus, by using the cohort approach to study actual fertility, one gets a similar picture, albeit a more accurate one, to that given by the crude birth statistics. Fertility declined for many years after the initial white contact. Then, as the percentage numbers of mixed-bloods increased, and Aborigines adjusted to a changed way of life, fertility rose to unprecedented levels. Then slowly it began to decline again, although it remained considerably higher than the fertility of the average Australian woman.

The Australian census data do not permit a comparable cohort analysis of the fertility of the general community. To make a comparison between Aboriginal and Australian overall fertility, it is necessary to work out age-specific fertility rates. These are calculated on the basis of numbers of children born per thousand women in certain age groups. In Figure 16, a comparison is made between the age-specific fertility rate of the Point McLeay Aboriginal women and that of Australian women as a whole. To arrive at the Aboriginal figure, a period of fifteen years, 1940-54, was taken and an average annual rate was calculated. This long period was necessary, because the figures for one or even for five years do not give a reasonable average when such small numbers are involved. A trial graph drawn on a five-year average was rather erratic and so a fifteen-year average was taken to reduce irregularities. The non-Aboriginal rate was obtained from the 1954 census data. This is not the most recent census but the one which fell within the fifteen-year period taken for the Aboriginal group.

It is evident that there are significant contrasts between the two populations of women. The Aboriginal fertility rate is higher than that of non-Aborigines in all age groups. The difference between the two is not as great in the optimum European age groups (ages twenty to thirty-four) as it is in the older age groups. European women tend to concentrate their child-bearing into a relatively short span of ten to fifteen years, but Aboriginal women bear children for a much longer period of time. Figure 16, then, agrees with Table 4 in suggesting that the main reason why Aboriginal women have a higher fertility rate than other Australian women is that they start bearing children when they are relatively young and continue right through to the menopause.

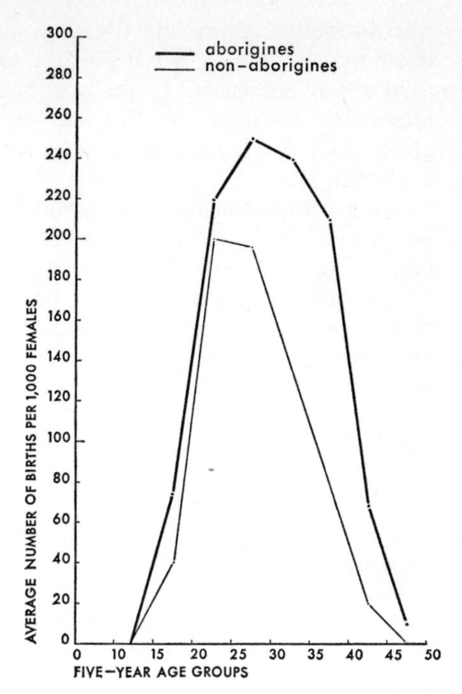


FIG. 16 Age-specific fertility rates of Aboriginal females, 1945-59, compared with those of other Australian females in 1954

## MORTALITY

The cohort approach is used in this paper to study changes in mortality as well as changes in fertility. As with the fertility section, fifteen-year birth cohorts are again used, but this time males as well as females are included. Table 5 shows the number of Aborigines in each cohort who survive at any particular age, in relation to the number born in each cohort. To facilitate comparison between the cohorts, the figures are worked out on the basis of the relative numbers of survivors per thousand births.

Since infant mortality has been rather high in the past, it would be preferable to separate deaths at birth from post-natal deaths. Furthermore, the period from birth to five years of age should be subdivided at least once, so that children who die before the age of one year are distinguished from those who die between the ages of one and five years. It is not possible, accurately, to make this distinction. Since the data are not obtained primarily from death registrations, it is not possible to be sure of the precise age at death in all cases. Births and deaths can be assigned, with reasonable accuracy, to five-year periods, but they cannot be given exact dates. Registered dates were available in many cases, but not in all.

More comprehensive data would be needed to separate natal, infant and child mortality. It is doubtful whether it will ever be possible to obtain data in sufficient detail to allow for such fine analysis of populations in past periods. So often an informant would say so-and-so 'lost a child', or refer to an offspring who 'died as a baby'. But it was not clear exactly how old the said child or baby might have been when it died. In many cases, further questioning revealed whether the baby had died at birth or some weeks later, but this could not always be verified. It is also possible that some still-born children may have been included in the infant mortality figures because Aborigines did not always distinguish between babies who were still-born and those who died at birth or soon afterwards.

Furthermore, it is likely that some recent deaths have been omitted unintentionally. Aborigines talk far more readily of births than of deaths. Although this research dealt with mixed-blood people who have lost most of the traditional reticence at speaking about a deceased person, they still retain a little of the sense of 'taboo' which veils such discussions. This applied more to recent deaths than to deaths well in the past. Probably most of the gaps have been filled in by the studies we made of social case files and especially of hospital records. Nevertheless, it is likely that mortality, especially in recent years, has been slightly higher than these figures suggest.

For all these reasons, the following mortality table is rather generalized. Precise figures may not be quite accurate, but the overall trends can be seen from the five-year age levels used. There is, of course, a drop off in the older age groups, especially with the later cohorts. Even so, quite definite trends can be seen. Infant and child mortality, under five years of age, have been high throughout the group's history. But there has been a marked improvement since the 1930s.

Infant and child mortality reached their highest levels for the cohort born between 1900 and 1914. It was in this period that the introduced European diseases took their greatest toll. There was no medical care available on the reserve at this time. Archival

TABLE 5

*Mortality*

| Birth Cohorts | Actual No. of Births | Number of Survivors per Thousand Births at the Following Ages: 5 yrs | 10 yrs | 15 yrs | 20 yrs | 25 yrs | 30 yrs | 35 yrs | 40 yrs | 45 yrs | 50 yrs | 55 yrs | 60 yrs | 65 yrs | 70 yrs | 75 yrs | 80 yrs |
|---|---|---|---|---|---|---|---|---|---|---|---|---|---|---|---|---|---|
| 1870–84 | 122 | 721 | 672 | 656 | 623 | 574 | 566 | 549 | 525 | 467 | 434 | 410 | 369 | 246 | 148 | 82 | 25 |
| 1885–99 | 174 | 730 | 648 | 638 | 569 | 477 | 425 | 402 | 374 | 322 | 276 | 247 | 201 | 155 | — | — | — |
| 1900–14 | 232 | 642 | 591 | 543 | 522 | 504 | 470 | 448 | 435 | 418 | 379 | — | — | — | — | — | — |
| 1915–29 | 256 | 676 | 641 | 625 | 617 | 605 | 584 | 570 | — | — | — | — | — | — | — | — | — |
| 1930–44 | 463 | 836 | 808 | 799 | 795 | — | — | — | — | — | — | — | — | — | — | — | — |
| 1945–59 | 630 | 881 | — | — | — | — | — | — | — | — | — | — | — | — | — | — | — |

material suggests that mortality was high when the mission was first set up in 1859. Then under the careful supervision and protection from contact with Europeans, mortality declined and population numbers gradually revived. This pattern is borne out by the figures which can be obtained for the cohorts born between 1870 and 1899. Infant and child mortality is lower for both these cohorts than it is for the next two cohorts. From about 1900 onwards, mission care dwindled and more and more Aborigines came into contact with Europeans. Increasing contact with Europeans and declining medical supervision apparently combined to give a quite significant rise in mortality rates after 1900. The decline in mortality since 1930 has been equally dramatic.

Thus, a two-phase cycle is evident. Infant and child mortality rose rapidly when the mission was first established and then declined towards the latter part of last century. At the end of last century and the beginning of this century we see a second rapid increase in infant and child mortality. Again, this is followed by a significant decline in recent years. There is no doubt that improved medical knowledge has been the cause of the more significant of the two examples of a decline in infant mortality—the decline in recent years.

## CONCLUSION

Certain demographic trends are evident from this study of the Point McLeay Aborigines. It is likely that similar trends apply in all those Aboriginal groups which encountered rapid contact with Europeans before the development of modern medical care. In the initial period of contact, the introduced diseases and a changed way of life resulted in a death-rate so high that the population declined. High infant mortality was a very significant factor in this decline. The period of time over which such a decline took place probably varied from one group to another, depending upon the rate and degree of European contact.

At Point McLeay there was a gradual introduction of both European diseases and culture from the time when the first sealing and whaling stations were set up on Kangaroo Island in the early part of the nineteenth century. Change was forced upon the Aborigines, and its rate accelerated when the first settlers arrived at Adelaide in 1836. The Adelaide Aboriginal group was rapidly decimated and its remnants were taken to Point McLeay to join other Aborigines who were being collected at this site after 1859. From this time onwards, the settled existence of Aborigines and their confinement on the mission gave the introduced diseases every opportunity to spread. The decline in the population was accelerated. By the end of the century, it was firmly believed that the group would die out altogether.

Approximately half a century after the settlement had been established, Aborigines were developing immunity to European diseases and were adjusting to the sedentary way of life. The group did not die out. At Point McLeay, and also, one may assume, at all of the older, established reserves, the mixed-blood birth-rate was increasing very rapidly. When linked with declining infant mortality, the outcome was actually a natural increase by the end of the nineteenth century. This process continued until there were too many people to be confined in such a small area: outward migration then began to increase. Those who moved away from the reserve and took up residence within the general Australian community limited their families to a size that conformed with that of their white neighbours. The overall birth-rate has thus declined, especially over the last two decades.

The 10 per cent (289 out of 2,625) which was excluded from this analysis because of insufficient data, comprised descendants of emigrants. No doubt they, too, have limited their family size in recent years. This is a logical assumption since this is the group that has managed to become completely assimilated into the general Australian community. Had it been possible to include these people in the analysis, the decline in the overall birth-rate of the last two decades, and, of course, in the total fertility rate, would probably have been more marked.

However, the birth-rate on the reserve has remained high. Thus, as long as Aborigines are encouraged to leave the reserves and become a part of the general community, the high fertility evident amongst mixed-bloods in earlier decades may continue to decline. But as long as reserves continue, it seems likely that those who live in such separated communities will continue to evince birth- and death-rates considerably higher than the Australian norm.

Where contact has been more recent and medical care more effective, as in Central and Northern Australia, the full-bloods have not died out. In the more isolated areas they have had time to adjust to a changed way of life and gradually to develop immunity to European diseases. In actual fact, the available data (Jones 1963) suggests that the numbers of full-bloods in the northern reserves are now increasing. It is yet to be seen whether they will follow the same pattern of increasing hybridization and rapidly rising birth-rates that the older reserves experienced at an earlier date.

What is clear, however, is that Aborigines who live in separate communities show very different demographic patterns from those who live within the general Australian community.

## ACKNOWLEDGEMENTS

It is not possible to acknowledge by name the directors and staff of the various government departments and hospitals whose collaboration made the study feasible. Miss Elaine Treagus, social worker, and Mrs Alison Brookman, research assistant, supported by the Australian Institute of Aboriginal Studies, assisted in the collection and checking of the genealogy. Miss Elizabeth Wittwer, tutor, Department of Geography, University of Adelaide, assisted with the analysis, and suggestions given by Mr Trevor Griffin, Lecturer of the same department, were most useful. Dr Frank Jones, Fellow in Sociology, Australian National University, read the draft script and gave invaluable help with the methods used in presentation. Finally, my thanks go to my husband who worked with me in writing the final paper. The magnitude of detail required in this study has virtually meant that it was a team project.

## BIBLIOGRAPHY

JONES, F. LANCASTER (1963). *A Demographic Survey of the Aboriginal Population of the Northern Territory, with special reference to Bathurst Island Mission.* Australian Institute of Aboriginal Studies, Occasional Paper No. 1, Canberra.

METGE, JOAN (1964). *A New Maori Migration.* London School of Economics Monographs on Social Anthropology No. 27, London.

PRICE, A. GRENFELL (1949). *White Settlers and Native Peoples.* Georgian House, Melbourne.

R. K. HEFFORD

# *Apartheid: Background, Problems and Prospects*

*'South Africans . . . face a task of appalling difficulty in reconciling the interests of the races.'*

A. GRENFELL PRICE *White Settlers in the Tropics*

## THE PLURAL SOCIETY

IN THE REPUBLIC of South Africa, the urban-rural distribution of settlers of European origin is now very similar to the distribution of population in other areas of relatively recent white settlement. But unlike the situation in Australia and North America, where the numbers of native peoples were sharply reduced by introduced diseases, induced malnutrition and deliberate slaughter, and have long since ceased to be of any political consequence, the whites of South Africa have always been greatly outnumbered by the indigenous peoples. Quite apart from the predominance of the Bantu, the whites in Natal, since the end of the nineteenth century, and in Cape Province over the last two decades, have also been outnumbered by peoples of other racial origin—predominantly Asiatics and the so-called 'Coloureds' respectively.[1]

For example, at the time of the census of 17 April 1904, the 97,109 whites in Natal constituted 8·8 per cent, Asiatics 9·1 per cent and the Bantu 81·5 per cent of the population. By the census of 8 May 1951, the whites had grown in numbers to 274,240 and accounted for 11·3 per cent of the population of Natal, the Asiatics 12·4 per cent and the Bantu 75·0 per cent. In Cape Province, at the time of the earlier census, the 579,729 whites constituted 24·1 per cent, Coloureds 16·6 per cent and the Bantu 59·1 per cent of population in that area. By 1951 the whites, 935,085 in number, accounted for 21·1 per cent, the Coloureds 22·2 per cent and the

[1] The Asiatics are almost exclusively of Indian origin: the Coloureds are very largely of mixed blood, but also include people of Malay and Griqua stock. Although the classification is ethnically unsatisfactory, the whites, Coloureds, Asiatics and Bantu—the latter being a collective term for 'Ngani, Sotho, Venda and Tonga-Shangana—will be regarded throughout as distinct racial groups, since it is on this basis that most official data are available and upon which the policy of apartheid has been founded.

Bantu 56·3 per cent of the population. Thus quite apart from the predominance of the Bantu, the whites in some areas have also been progressively outnumbered by either the Coloureds or the Asiatics.

The census of 6 September 1960 revealed that, in a total population of 15,994,181, the non-whites outnumbered whites by approximately four to one, although the proportion ranged from eight to one in Natal to three to one in the Transvaal; furthermore, that the Asiatics in Natal and the Coloureds in Cape Province were increasing in numbers much more rapidly than were the whites in either area.

TABLE 1

*Racial Composition of Population by States, September 1960*

(thousands and per cent of total)

| | Whites | | Coloureds | | Asiatics | | Bantu | |
|---|---|---|---|---|---|---|---|---|
| Natal | 337 | (11·3) | 45 | (1·5) | 395 | (13·3) | 2,200 | (73·9) |
| Cape Province | 1,001 | (18·7) | 1,330 | (24·8) | 18 | (0·3) | 3,011 | (56·2) |
| Transvaal | 1,466 | (23·4) | 108 | (1·7) | 64 | (1·0) | 4,633 | (73·9) |
| Orange Free State | 276 | (19·9) | 26 | (1·9) | — | (—) | 1,084 | (78·2) |
| Total | 3,080 | (19·3) | 1,509 | (9·4) | 477 | (3·0) | 10,928 | (68·3) |

*Source:* Republic of South Africa, Bureau of Statistics: *Population Census, 6 September 1960.*

The composition of population in the cities, however, presented a marked contrast to the position in the Republic as a whole. In most of the major cities, non-whites outnumbered whites by only two to one and in some cities, notably Pretoria (1·04 to 1), but also Cape Town and Johannesburg, by an even smaller margin.

TABLE 2

*Population (Thousands) and Percentage Composition by Races in the Three Principal Cities, September 1960*

| | Population | Whites | Coloureds | Asiatics | Bantu |
|---|---|---|---|---|---|
| Johannesburg | 1,149 | 35·8 | 5·2 | 2·5 | 56·5 |
| Cape Town | 806 | 37·8 | 51·8 | 1·1 | 9·3 |
| Durban | 681 | 28·6 | 4·0 | 34·8 | 32·6 |
| Republic | 3,080 | 19·3 | 9·4 | 3·0 | 68·3 |

*Source:* Republic of South Africa, Bureau of Statistics: *Population Census, 6 September 1960.*

Although Table 2 clearly indicates that the whites are concentrated in the cities, the extent of concentration is understated: the data, in each case, include population in nearby rural areas and therefore fail to reveal fully the effects of deliberate diversion of non-whites from city centres to urban fringes or rural areas.[2] Nevertheless, it is clear that the cities are the real centres of white supremacy.

As a corollary to this greater concentration of whites in the principal cities it follows, of course, that the whites were relatively fewer in numbers and the Bantu much more heavily concentrated in rural areas. In fact, while the 1960 census classified only 16·4 per cent of the whites, 16·8 per cent of the Asiatics and 31·7 per cent of the Coloureds as resident in rural areas, 68·2 per cent of the more numerous Bantu were so classified. Thus non-whites outnumbered whites by sixteen to one and Bantu outnumbered all other races by seven to one in such areas.

Vital statistics available for 1965 show, for the whites, a crude birth-rate (CBR) of 22·8 per thousand and a crude death-rate (CDR) of 9·1 per thousand, resulting in a natural increase equivalent to 1·36 per cent per annum (Republic of South Africa: *Statistical Year Book, 1966*). Amongst the non-whites, CBR's varied between races but were of an order generally associated with under-developed countries: 34·4 and 46·1 respectively for the Asiatics and Coloureds and, in the absence of published data and on the basis of age composition only, 35-40 amongst the Bantu. CDR's for the Asiatics and Coloureds were 8·1 and 15·8 respectively, and for the Bantu probably 13-18 per thousand, in each case reflecting the substantial impact of modern medicine. The rates of natural increase amongst the non-whites were therefore 3·03, 2·63 and an estimated 2·25 per cent per annum for the Coloureds, Asiatics and Bantu respectively.

Regarding the future population of South Africa, these vital statistics alone suggest that numbers in each non-white group will grow more rapidly than the numbers of whites. In fact, the downward trend in white CBR, from more than 25 per thousand ten years ago, suggests that the recent rate of natural increase might

[2] Throughout the 1960 census (*vide* Preface to Vol. 2, No. 5), 'a metropolitan area is generally defined as an urban complex consisting of a parent municipality, together with the adjoining areas which are urban in character and which are economically and socially linked with the parent town'. The data in Table 2 therefore refer, in each case, to several magisterial districts (or statistical subdivisions) and cover areas ranging from 800 to 1,200 square miles.

not be sustained. In contrast, the Coloured CBR and CDR have remained fairly constant over the past decade, infant mortality largely accounting for fluctuations in natural increase; the Asiatic CBR has been rising steadily since the early sixties while CDR, primarily as a consequence of declining infant mortality, has fallen slightly, resulting in an increase in growth rate of 0·86 per cent per annum since 1957. Any further improvement in the welfare of the Bantu will no doubt lead to a decline in CDR and a corresponding increase in the population growth rate.

The numbers of whites might, of course, be augmented by immigration, and it is evident from alternative official projections of population growth—in which estimates of future white population are based on assumed growth rates ranging from almost 1·9 to 2·25 per cent per annum—that the government of South Africa, with or without foundation, is anticipating substantial net immigration.

TABLE 3

*Population Projections: Median Assumptions* [3]

(expected population by races at 30 June: thousands and % of population)

| | 1965 | | 1975 | | 1985 | | Assumed Annual Growth rates (%) |
|---|---|---|---|---|---|---|---|
| Whites | 3,398 | 19·0 | 4,155 | 18·5 | 4,992 | 17·8 | 2·0 |
| Coloureds | 1,751 | 9·8 | 2,377 | 10·6 | 3,235 | 11·5 | 3·1 |
| Asiatics | 533 | 3·0 | 672 | 3·0 | 841 | 3·0 | 2·3 |
| Bantu | 12,186 | 68·2 | 15,225 | 67·9 | 19,020 | 67·7 | 2·3 |
| Total | 17,868 | | 22,429 | | 28,088 | | |

*Source:* Republic of South Africa: *Statistical Year Book, 1966.*

Despite an unrealistically low growth rate assumed in respect of the Asiatic population, which might have been expected to increase at 2·7 or 2·8 per cent per annum in view of recent trends, a growth rate for the Bantu which appears to make no allowance for improvements in health and education services on the level of mortality, and possibly heroic assumptions regarding future numbers of white immigrants, these projections nevertheless foreshadow a further increase in the ratio of non-whites to whites in the Republic.

[3] Population for 1965 is a mid-year estimate based on re-evaluation of the September 1960 census. Although the basic assumptions underlying these projections were not specified, other official sources indicate that the government of the Republic anticipates future net white immigration at the rate of approximately 30,000 per annum.

## THE EVOLUTION OF APARTHEID

The whites of South Africa have now lived for many generations as a minority racial group and throughout this period, not surprisingly, they have clung tenaciously to many aspects of the cultures imported formerly by their Dutch and English forefathers. Although the assimilation of migrants has generally progressed rapidly after the second generation in other areas of comparatively recent settlement, resistance to assimilation is not, of course, unique. The majority of Chinese in South-east Asia, for example, have not only continued to resist assimilation but also to indoctrinate their children with aspects of the 'homeland' culture. Such attitudes are clearly a manifestation of nationalism.

In the South African case, white nationalism in similar form has long been in evidence, those of Dutch origin being equally or more determined to maintain the culture and traditions of Holland than to simply share with those of English origin in the maintenance of a European civilization. In fact, the Afrikaner *volk* have long sought completely *separate development* (apartheid) and it was this uncompromising Afrikaner nationalism, antagonistic to integration with the English and apprehensive lest their own culture be obliterated, that led finally to the so-called Boer War. Nor did subsequent self-government, bestowed on the Union in 1909, or the community of interest necessarily imposed by participation in two world wars, serve to stem the extremist element in Afrikaner ideology.

In order to appreciate the motivating forces behind white nationalism, it should be remembered also that most whites came to South Africa as permanent settlers and not, as did the majority of their counterparts in East Africa and South-east Asia, as 'temporary sojourners'. The approach to settlement in South Africa has long been remarkably similar to that elsewhere in the 'new world'. Investment, including heavy and extensive social overhead expenditure, has been undertaken in every conceivable direction, in marked contrast to investment almost exclusively in direct support of plantations, mines and trading posts in many of the former tropical colonies. The fact remains that the whites of South Africa have for many generations sought to build for their children a society such as exists in other areas of relatively recent but permanent white settlement: the only notable distinction, in the South African case, lies in the fact that the whites there have always been heavily outnumbered by coloured peoples.

Having created for themselves such a substantial area of what they had come to regard as permanent settlement, it was inevitable that the first (wartime and immediate post-war) rumblings of discontent in the African and other colonies should have created an uneasiness in the minds of white South Africans. It was not difficult, from promises issuing from the metropolitan powers, to visualize the shape of things to come. In particular, promises of or pressures for ultimate self-government made clear the very tentative foothold held by white settlers in colonies such as Kenya, Uganda, Tanganyika and Nyasaland: in these four colonies, as late as 1958, the Europeans collectively numbered only 104,100 in a combined population of 23,772,000. In contrast, the whites in South Africa at that time numbered very nearly three millions in a population of a little more than fifteen millions.

Although it was clear, from 'the winds of change', that the white minorities in most African colonies would soon be forced to leave, it is not surprising—in fact, understandable—that the white South Africans were determined to remain. Nor is it surprising that the whites of South Africa, either fervently nurturing or frequently thrust into contact with the belief that in the separation of the races lay the only path to survival for each culture, at this time saw in the doctrine of apartheid some hope for the future.

It is not generally appreciated that apartheid, formerly a rather nebulous concept, took on concrete form with quite explicit objectives twenty years ago. Immediately prior to its General Election victory of May 1948, the National Party issued a statement in which it asserted that the whites of South Africa must choose 'one of two directions. Either we must follow the course of equality, which must eventually mean national suicide for the white race, or we must take the course of separation (apartheid) through which the character and the future of every race will be protected . . .' (Malan 1948). Thereafter followed a detailed enumeration of the policies which the National Party, if elected, proposed to put into effect: for example, 'All marriages between Europeans and non-Europeans will be prohibited. . . . The party wishes all non-Europeans to be strongly encouraged to make the Christian religion the basis of their lives. . . . All Natives must be placed in separate residential areas . . . and . . . can never be entitled to any political or equal social rights . . . in European areas. . . . The party is opposed to the organisation of Natives into trade unions, and advocates a system whereby the State, as guardians, will take care of their interests. . . .'

Dr D. F. Malan, in a pre-election speech at Paarl in Cape Province on 20 April 1948, and Paul Sauer (then Chief Whip in the National Party), in an article in the *Cape Times* on 30 April 1947, clarified other aspects of the proposed policies, particularly as relating to the individual non-white races (Malan 1948). Sauer suggested that the Coloured people 'have culturally, economically and geographically thrown in their lot with the European . . .'; Dr Malan, that 'they share the same language and cultural interests and . . . must hold a privileged position in the European areas'. Both emphasized that the Coloureds were to be accorded preferential treatment, both in employment and in establishing business ventures, over other non-whites.

In contrast, and concerning the Bantu, Sauer stated that 'the Natives are in a somewhat different category . . . even the detribalised . . . Native has often little more than the superficial veneer of European civilisation.' Although he went on to pledge that the National Party would assist in Bantu development, no indications were given regarding the manner in which such assistance was to be provided. Sauer merely stipulated that the Bantu must achieve development largely in the isolation of their reserves: 'The Native areas must be treated as his true home, . . . and there . . . he must learn to become self-sufficient. . . . In European areas he must be considered a temporary worker. . . .'

Dr Malan's attitude at this time towards the Indians in South Africa was harsh indeed—particularly in view of their not insignificant role in the economy, and the realization that they, like their counterparts in East Africa, held no political aspirations. Three statements culled from Dr Malan's speech at Paarl appear sufficient to clarify the attitude taken: '. . . the Party will strive to repatriate or remove elsewhere as many Indians as possible'; 'the Cape urban areas must also be protected against Indian penetration' and 'family allowances to Indians must be abolished'.

Overall, and notwithstanding the preferred (but second-best) position of the Coloureds, both Malan and Sauer emphasized that the principle of separateness must be applied to all non-whites with regard to places of residence, means of public transport, in recreation and, as far as possible, in places of work.

In the matter of political representation, Sauer admitted that 'the European fears the political power the non-European might have or develop': he suggested that the remedy lay in separate constituencies and electoral rolls. But Dr Malan, speaking a year later, put the matter of future political representation beyond any doubt: 'In view of the possession of their own national home in the reserves, Natives in the European areas can make no claim to political rights. The present representation of Natives in Parliament and in the Cape Provincial Council must therefore be abolished'; and, regarding the Indian community, 'no representation will be given to Indians in the legislative bodies of the Country'. Thereafter, the Coloureds were to be represented by one senator nominated by the government and three elected by the Cape Provincial Council, the Bantu by three senators appointed by the government and four elected by native councils—but with the significant provisos that all such senators must be of European origin and none would be permitted to vote on various matters specified and including, *inter alia*, the political rights of non-Europeans.

This, then, was the substance of the apartheid platform of the National Party when it received its mandate from the electorate in May 1948 and, apart from some hardening in attitude towards the Coloureds and failure to implement some sanctions originally proposed against the Indians, these are the policies that have since been pursued. Meanwhile, and despite world opinion, the white South Africans have continued to return the National Party to office, with support for apartheid stemming largely from the assumption that white supremacy in South Africa serves to limit Communist penetration in Africa—a feeling equally strong amongst supporters of the Opposition Party—but also from a belief that Bantu welfare is, and will continue to be, substantially above that enjoyed by peoples in the independent states of black Africa. Certainly repeated promises by Sir de Villiers Graaf that his United Party would dismantle some of this discriminatory apparatus have not won support from the white electorate.

Charles Manning, a South African by birth, has suggested that 'what underlies apartheid is at bottom an attitude not towards the black man but towards the forefathers . . . of the Africaner people. . . .' (Manning 1964). Prior to the time of 'the winds of change', this could have been accepted as a reasonable prognosis. A subsequent statement by Manning seems more relevant to the current issues: 'Deplore the white man's collective self-concern, and you may equally well damn every other example of nationalism, white or black. It is absurd to assume that nationalism is nice, or nasty, according to its colour.' It is true enough that coloured nationalism, if not always understanding or understood, now carries with it an air of respectability, whereas white nationalism in any form—as in the restrictions recently imposed on the flow of Asian migrants from Kenya into Britain—is invariably condemned. But in the South African case, it is not white nationalism *per se* that the world condemns. Nor does a consensus of world opinion suggest that, as in East Africa, the white South Africans should simply abdicate in favour of the black majority: the widespread censure is generally based on social aspects of racially discriminatory policies and refusal of the white minority to grant any meaningful political representation to coloured peoples.

## THE SOCIAL AND ECONOMIC CONSEQUENCES OF APARTHEID

The social consequences of South Africa's racial policies are now widely known and readily apparent to the most casual observer. Non-whites have been removed from the city centres—the conspicuous exception being the Indians of Durban—and these former residential areas have been generally demolished to make way for impressive public buildings or blocks of flats reserved exclusively for white use. Separate public conveniences, separate entrances to public buildings, separate means of public transport and even separate areas for bathing at the beaches are clearly marked 'Europeans Only', 'Whites Only' or 'Non-whites Only'—all of which seems quite ludicrous in view of the extent of social intercourse necessarily involved in most factories and other places of work. More recently, as reported in the *Cape Argus* on 20 January 1968, and to take effect from the end of that month, 'Coloured bands are not allowed to provide music for the entertainment of Whites in a White-proclaimed area.'

Such measures clearly suggest that skins other than white are a mark of inferiority. Discussion with South African whites reveals attitudes ranging from guarded tolerance to complete distrust and intense antagonism, particularly towards the Bantu. Reference is commonly made to excessive drinking, illegitimacy and other forms of immoral conduct although, in some instances, the concept of 'immorality' is apparently based largely on the traditional Bantu exchange of women for cattle. Emphasis is also placed on non-white propensity to physical violence. In order to avoid the possibility of attack, white drivers are advised by police to drive on after any accident involving non-whites. Shop windows in some areas are heavily meshed against the threat of robbery and truncheon-carrying police, many of them non-white, are very much in evidence in city streets at night.[4] There is clearly evident in the white community a climate of fear, perhaps with substantial foundation in view of the recent report (Cape Edition of *Post*, 21 January 1968) that Groote Schuur Hospital alone treated more than five thousand stabbing cases in 1967. Whether and to what extent such violence can be attributed to racialist policies is difficult to determine.

Discrimination is also evident in the forms of employment open to non-whites in the face of legislation reserving specified occupations exclusively to whites. Bantu in urban areas, for example, are rarely seen in anything other than unskilled and semi-skilled occupations such as road and building construction, the handling of freight, vehicle operation and various forms of domestic service. Examination of employment opportunities advertised in the newspapers of Johannesburg, Cape Town and Durban leads to similar conclusions. Employers invariably specify that they seek 'Europeans' for skilled and administrative positions, and 'Indians' or 'Indians and Coloureds' for clerical work, truck driving or machine operation in manufacturing industry. Apart from occasional opportunities for employment as labourers in manufacturing industry or as truck drivers, Bantu are rarely sought for any position other than as domestic help. (These impressions are readily confirmed by reference to official statistics: see Table 6.)

Meanwhile, people continue to be imprisoned or deported for marriages in contravention of the state colour bars, African unions are still denied official recognition and, in Feburary 1968, legislation was introduced to remove the last pretence of non-white representation in the Parliament of the Republic. The recent imposition of a ban on the Moslem call to prayer in a village in Western Transvaal, allegedly to remove a source of 'public nuisance', is but another example of a lack of respect for fundamental human rights (Johannesburg *Sunday Express*, 21 January 1968).

Clearly, the pursuit of such policies cannot be reconciled with the Charter of the United Nations, Article 1 of which calls for, *inter alia*, 'respect for human rights and for fundamental freedoms for all without distinction as to race, sex, language or religion. . . .' Thus the whites of South Africa should not have been surprised

[4] Non-white police are not, of course, permitted to arrest whites. Nor is L. H. D. Mouli, the first African magistrate appointed to the administration of Bantu areas (effective from 1 April 1968) empowered to try whites in his courts, notwithstanding that he is academically better qualified than many white magistrates in the Republic.

by the nature or extent of criticism offered in recent years. In view of both their origins and the strongly-held belief in their role in the struggle against Communism, they apparently expected some sympathy from Western countries. But from such sources has come some of the most damning condemnation, such as in the following response from Philip Mason (former Director of Studies in Race Relations, Royal Institute of International Affairs): 'It would not be easy to judge whether the right to security of person—that is, freedom from torture and arbitrary arrest—is respected any more or any less in South Africa than in one of the Iron Curtain countries. But equality before the law, freedom of movement, freedom to marry, are clearly much less evident in South Africa. Catholics in Poland may not have the right to educate their children as they want, but they are not debarred from living with their wives and children . . . nor forced if they lose a job to give up their houses and move to some remote part of the country. They are not confronted with benches on which they may or may not sit and vehicles in which they may not travel. They are not sent to segregated schools and forbidden to marry outside their community . . .' (Mason 1964).

Numerous additional examples of criticism of apartheid could be quoted, each emphasizing the sanctity of freedom in its most fundamental forms. But little attention has been given to claims made by the supporters of apartheid regarding the material welfare of non-whites in South Africa. If, as some white South Africans suggest, Bantu welfare is far in excess of the levels available or likely to be experienced in black Africa, the implied corollary appears to be that—if, indeed, they had any choice in the matter—the choice for the Bantu would lie between freedom in the midst of poverty and tolerable material welfare in a community where fundamental freedoms are denied. It is conceivable, of course, that the Bantu might prefer self-determination at any price. Nevertheless, the arguments in defence of apartheid still suggest the need for an evaluation of material welfare currently available to these people.

At the time of the 1960 census, 61·6 per cent of the whites but only 12·7 per cent of the Coloureds, 16·5 per cent of the Asiatics and 5·2 per cent of the Bantu had ever received a formal education to late primary or higher level. Of the respective populations of five years of age or more at the time of the census, only 5·1 per cent of the whites but 52·6 per cent of the Bantu had received no formal education whatever.

TABLE 4

*Highest Standard of Education Received to the Time of Census (6 September 1960)*

(percentage of population by race)

| Standard | Whites | Coloureds | Asiatics | Bantu |
|---|---|---|---|---|
| Std 8 or higher | 35·1 | 2·8 | 4·5 | 1·2 |
| Std 6 or 7 | 26·5 | 9·9 | 12·0 | 4·0 |
| Std 3, 4 or 5 | 10·7 | 21·4 | 23·6 | 11·6 |
| Std 2 or lower | 8·6 | 18·0 | 19·8 | 14·9 |
| Nil | 16·5 | 47·1 | 39·1 | 68·1 |
| Not specified | 2·6 | 0·8 | 1·0 | 0·2 |

*Source:* Republic of South Africa: *Statistical Year Book, 1966.*

By 1963, the gap in educational attainment between Asiatics and whites was rapidly closing, with 46·9 per cent of the estimated (and younger) Asiatic population at school, compared with 44·9 per cent of the whites, and a marked increase in the proportion of Asiatics engaged at secondary and tertiary levels. But the proportions of Coloureds and Bantu undergoing formal education (33·3 and 20·4 per cent respectively), had shown no improvement except for a very slight increase in the proportions proceeding to secondary or tertiary level. Nevertheless, the numbers of Coloured and Bantu students now receiving a tertiary education are pitifully low in relation to their respective populations: in 1965, for example, there were 42,918 full- or part-time students in the universities and university colleges of the Republic, of whom 39,392 were white, 723 Coloured, 1,721 Asiatic and 1,082 Bantu.

Were it not for 'job reservations' and non-recognition of African unions, it would be tempting to seek to explain employment opportunities available and wage rates paid to non-whites largely in terms of scholastic achievement. But the distribution of the workforce between occupations, particularly the proportions of each workforce engaged in clerical, sales and production-line occupations, indicates that something more than educational attainment is involved.

The whites in South Africa are distributed between industries in a manner generally expected in an industrialized economy. The proportion in agriculture, for example, is very similar to the proportions so engaged in the United States and Australia. The proportion in mining is, understandably, somewhat more than is generally the case; the proportions in commerce and personal

TABLE 5

*Percentage of Workforce by race and by Industry, September 1960*

| | Whites | Coloureds | Asiatics | Bantu |
|---|---|---|---|---|
| Agric/forestry/fisheries | 10·4 | 21·8 | 8·6 | 37·0 |
| Mining/quarrying | 5·4 | 0·8 | 0·5 | 14·1 |
| Manufacturing | 18·2 | 16·8 | 25·2 | 7·9 |
| Construction | 6·3 | 7·2 | 1·8 | 4·1 |
| Electricity/water/gas | 0·9 | 0·5 | 0·2 | 0·7 |
| Commerce/finance | 22·2 | 8·0 | 23·0 | 4·9 |
| Transp/communications | 10·1 | 3·0 | 3·0 | 1·8 |
| Gov't/personal services | 22·6 | 26·2 | 18·7 | 21·1 |
| (of which government | (7·6) | (2·1) | (2·0) | (1·8) |
| health/ education | (7·2) | (3·1) | (4·4) | (1·8) |
| pers. services) | (3·0) | (20·0) | (10·5) | (16·7) |
| Unemployed/unspecified | 3·9 | 15·7 | 19·0 | 8·4 |

*Source:* Republic of South Africa: *Statistical Year Book, 1966.*

services are somewhat less, but the numbers of Indians engaged in commerce and Coloureds and Bantu performing personal services more than compensate for any apparent deficiencies.

In the Republic as a whole, the Coloureds are concentrated very heavily in agriculture, manufacturing and personal services, these three industries together accounting for almost 70 per cent of the gainfully employed at the time of the 1960 census. In the statistical divisions covering each of the three principal cities, the proportion of Coloureds engaged in agriculture is naturally very much less than the national average and the proportion engaged in manufacturing much greater: in Johannesburg, for example, a little more than half the Coloureds employed at the time of the census were engaged in manufacturing. On the other hand, as must be expected in an economy encumbered by a surfeit of unskilled labour, the unemployed—or those officially permitted to do so—tend to drift to the cities in search of employment: thus one-fifth to one-quarter of the Coloured workforce in the Johannesburg, Cape Town and Durban subdivisions was unemployed at the time of the census, compared with less than 5 per cent of the whites in each of these areas.

The Asiatics are not as heavily dependent on commerce as first impressions of South African cities would suggest although, in cities such as Johannesburg and Cape Town, where their numbers are relatively few, the fields of commerce and finance absorb half or slightly more of the workforce. But in Durban, the stronghold of expatriate Indians, the scope for commerce is not unlimited and imposes the ultimate restriction on such job opportunities: whereas 15·2 per cent were engaged in commerce in 1960, 31·2 per cent were employed in manufacturing, 10·5 per cent in personal services and 22·8 per cent were unemployed. At the same time and throughout the Republic, the Asiatics were conspicuously absent from industries generally associated with harder physical labour—such as construction, mining and quarrying.

In contrast, the Bantu are much more heavily dependent on agriculture, mining and quarrying, these industries together accounting for 51·1 per cent of the workforce in 1960. Employment in the principal industries of urban areas (namely, manufacturing and commerce) together accounted for only 12·8 per cent of the Bantu, compared with 40·4 per cent of the whites, 24·8 per cent of the Coloureds and 48·2 per cent of the Asiatics. In fact, were it not for employment in personal services, accounting for up to one-third of the workforce in principal metropolitan subdivisions, there would be relatively few Bantu in urban areas.

Although a useful guide to the urban-rural distribution of opportunities, employment by industries does not indicate the levels at which the various races are employed. Classification of workers by occupation is much more revealing with regard to skills applied

TABLE 6

*Percentage of Workforce by Race and by Occupation, September 1960*

| | Whites | Coloureds | Asiatics | Bantu |
|---|---|---|---|---|
| Professional-technical | 12·1 | 2·5 | 4·1 | 1·2 |
| (of which nursing/ teaching) | (5·3) | (2·3) | (3·4) | (1·0) |
| Administrative-managerial | 5·1 | 0·3 | 1·9 | 0·1 |
| Mining tech'n-supervisor | 2·7 | 0·1 | — | — |
| Clerical | 24·2 | 1·6 | 6·5 | 0·5 |
| Salesworker | 8·6 | 1·9 | 18·3 | 0·7 |
| Transport-comm. worker | 5·5 | 3·9 | 5·7 | 1·6 |
| Craftsman-labourer | 24·0 | 34·7 | 28·7 | 32·3 |
| (of which labourer) | (1·5) | (17·1) | (7·7) | (30·1) |
| Service worker | 5·1 | 21·3 | 11·6 | 18·3 |
| (of which domestic) | (1·2) | (18·5) | (7·9) | (15·7) |
| Farm-fish-lumber worker | 10·3 | 23·0 | 9·2 | 38·0 |
| (of which farm labourer) | (1·1) | (21·2) | (6·2) | (25·8) |
| Unemployed-unspecified | 2·4 | 10·7 | 14·0 | 7·3 |

*Source:* Republic of South Africa: *Statistical Year Book, 1966.*

by the various races, if not necessarily, in view of the system of 'job reservations', indicative of the skills available.

Of 10·1 per cent of the white workforce engaged in farming at that time, almost nine-tenths had their own farms or market gardens: in contrast, 21·2 per cent of the Coloureds and 25·8 per cent of the Bantu—in the latter case alone involving 1,004,645 persons—were employed as farm labourers. In mining and quarrying, the whites participated almost exclusively as technicians and supervisors while 535,842 or all but 0·3 per cent of the Bantu indicated in Table 5 as engaged in mining and quarrying were, in fact, employed as labourers. Of the 34·7 per cent of Coloureds and 32·3 per cent of Bantu classified as 'craftsman, production worker or labourer'—a category, in the main, covering workers in manufacturing industry—16·6 per cent and 16·3 per cent (or 633,648) respectively were labourers in industries other than mining. In the 'Service worker' category (including police, caterers and people associated with recreational activities or laundries) 18·5 per cent of the Coloureds and 15·7 per cent (or 611,060) of the Bantu were employed as caretakers or domestic servants.

Thus 56·8 per cent of the Coloured and 71·6 per cent (or 2,785,195) of the Bantu workforce were employed either as labourers or domestics, occupations which obviously require little in the way of skills which might serve to further 'separate development'. (If those unemployed or who failed to specify occupation are added, this accounts for 67·5 per cent and 78·9 per cent respectively of the Coloured and Bantu workforces.) In contrast, the proportions of non-whites engaged in professional, technical, administrative and managerial occupations—particularly if nurses and teachers are excluded—are far less than necessary to the advancement of their own communities. The Bantu, in particular, are conspicuously absent from such occupations, indicating that the élite which South Africa is allegedly attempting to create amongst these people is as yet very thin indeed.

On the other hand, employers now tend to favour Indians ahead of other non-whites. This is particularly evident in Durban, where considerable ill-feeling exists between the Indians and Bantu—the latter claiming the Indians are filling occupations not rightly theirs. Whether or not the Bantu have the capital or the skills to match the Indians, Table 6 reveals a much greater proportion of the latter workforce engaged in more highly paid occupations. In fact, when compared with other races, the Indians tend to be unique only in that a greater proportion of their workforce is engaged as shop proprietors or assistants and a smaller proportion in each form of primary industry.

There are a number of factors bearing on income accruing to each racial group in South Africa, including age composition, proportion of each population gainfully employed, 'job reservations' and other measures influencing urban-rural distribution of population and forms of employment available, past opportunities for formal education, discriminatory attitudes of individual employers and, not least, wage rates that vary widely according to the racial origins of people engaged in identical work.[5]

From data relating to numbers of whites, Coloureds and Asiatics in major occupational categories and earning income within specified class intervals during 1960, it was estimated that average annual personal income earned by those employed in that period was as follows: white males R2,149, Asiatic males R640 and Coloured males R367; white females R904, Asiatic females R351 and Coloured females R259.[6] Since the proportion of whites in receipt of income over this period, 42·7 per cent, was far in excess of the proportion of Coloureds (34·6 per cent) and Asiatics (23·5 per cent), the distribution of aggregate personal income over the populations of each race revealed even greater inequality: average per capita income for the whites was R750, the Coloureds R109 and Asiatics R141, equivalent, at the then current exchange rate, to $U.S. 1,046, $U.S. 152 and $U.S. 197 respectively.

Unfortunately, no official data relating to Bantu income were available for that period. But as compared with the Coloureds, a smaller proportion (1·1 per cent less) of the Bantu population was gainfully employed; a much greater proportion of the Coloured workforce was employed in more highly paid occupations, particularly as craftsmen in manufacturing industry (see Table 6); and even in identical occupations, the Coloureds more often than not drew slightly higher wages. This, together with estimates based on various semi-official sources, suggests that average per capita income amongst the Bantu would have been appreciably lower

[5] For example, in 1962, average weekly earnings of motor vehicle drivers in Witwatersrand were as follows: whites, Rand 29-70; Coloureds, Rand 19-75; and Bantu, Rand 15-50. In Durban in the same year, Asiatic drivers received an average of Rand 17-90 and Bantu drivers Rand 13-23.

[6] The Rand, in that year, was equivalent to $U.S. 1.40. Of the whites, 7·6 per cent of the males and 25·7 per cent of the females earned less than R400; of the Asiatics, 45·1 per cent of the males and 73·7 per cent of the females earned less than R400; but of the Coloureds, 69·6 per cent of the males and 97·5 per cent of the females earned an annual income of less than R400 (*Statistical Year Book, 1966*).

than the R109 ($U.S. 152) of the Coloureds and possibly as low as R80 ($U.S. 112).[7] In fact, this latter estimate appears to have been confirmed by a recent official report which, emphasizing growth in Bantu income in recent years, placed average income at the equivalent of $U.S. 128 per capita in 1967.

Estimates of income in the less-developed and predominantly subsistence economies of Africa and elsewhere have long been regarded as extremely suspect, if for no other reason than the difficulties involved in assessing quantities and value of foodstuffs consumed by cultivators and their families. Many would agree with Frederick Benham that, notwithstanding the low $U.S. equivalent of national income statistics issued by some of these countries, real per capita income consistent with the bare maintenance of life in 1960 could nowhere have been less than the equivalent of $U.S. 100. Thus, while recognizing the need for caution, estimates of per capita income in Zambia and Ghana of $U.S. 170-190 in recent years seem quite feasible. In contrast, and bearing in mind that three-quarters of the Bantu necessarily live outside the reserves and are therefore unable to supplement cash income by recourse to subsistence crops, the estimates of Bantu income suggest living standards little, if anything, above what Benham would regard as minimal.

This inequality in income is reflected, in part at least, in the incidence of disease, infant mortality and average expectation of life.

TABLE 7

*Disease, Infant Mortality and Life Expectation by Race (1965)*

| | Whites | Coloureds | Asiatics | Bantu |
|---|---|---|---|---|
| Registered cases of notifiable disease (per 100,000 population) | | | | |
| Diphtheria | 5·1 | 9·8 | 6·0 | 10·4 |
| Malaria | 1·0 | 0·2 | 0·6 | 0·7 |
| Typhoid fever | 3·2 | 10·5 | 7·1 | 37·6 |
| Tuberculosis | 37·1 | 517·9 | 248·2 | 459·3 |
| Kwashiorkor | 0·3 | 42·3 | 5·0 | 99·6 |
| Infant mortality (per '000 live births) | 29·0 | 136·1 | 56·1 | ? |
| Expectation of life at birth (years) | | | | |
| Males | 64·6 | 44·8 | 55·8 | ? |
| Females | 70·1 | 47·8 | 54·8 | ? |

*Source:* Republic of South Africa: *Statistical Year Book, 1966.*

[7] Admittedly, some Bantu qualify for Social Security benefits, principally old age pensions and disability grants. But in 1965, such benefits in all forms added the equivalent of only $U.S. 0.98 to per capita income.

No official data are available regarding either infant mortality or expectation of life amongst the Bantu, but Philip Mason has suggested that infant mortality is 'as high as 400 per thousand in some rural areas'. Age composition of the Bantu suggests that life expectation would be slightly less than amongst the Coloureds. But official estimates of life expectation continue to be based on Life Tables computed for the period 1950-52 and in view of marked progress since achieved against some diseases, particularly diphtheria, cerebro-meningitis, infective encephalitis, malaria and kwashiorkor, life expectation should now be somewhat higher than indicated.

Although the evidence available is far from conclusive, the numbers of whites and non-whites convicted in South African courts of law are difficult to reconcile with the respective populations and suggest discrimination, particularly against the Coloureds. Of the massive total of 383,477 convictions for all offences during the year 1963-64, only 10·8 per cent were recorded against whites and 2·3 per cent against Asiatics, but 97,251 (or 25·4 per cent) against the Coloureds. As at 30 June 1965, 55 of the 59 being held under sentence of death were Coloured or Bantu; of 412 undergoing life imprisonment, all but 17 were Coloured or Bantu (*Statistical Year Book, 1966*). Is the higher rate of conviction amongst the Coloureds and Bantu due to living standards so low as to encourage crime? Is it that, as some white South Africans are quick to suggest, these races are particularly given to lawlessness? Or does the white administration simply commit potential dissidents to prison on the slightest pretext? Whatever the reason, the black peoples of Africa believe that evidence of discrimination exists and have repeatedly rejected white South African assurances of equality before the law.

## THE FEASIBILITY OF 'SEPARATE DEVELOPMENT'

Historically, the process of economic development in Western countries typically rested heavily on the entrepreneurial initiative of a wealthy minority responsive to trading opportunities at home and abroad, but backed by a progressively more productive and less labour-intensive farm sector that concurrently released ever-increasing supplies of both labour and foodstuffs in support of a leading or 'modern' sector.

Basically the same path must be followed by the currently under-developed countries. But most of these economies face formidable obstacles, not least of which are a generally low level of income and savings, the virtual absence of an élite capable of sound administration or willing to invest, a dearth of technical skills, 'bottlenecks' in the form of inadequate power supplies, transport services and other social overheads necessary to the foundation of a modern sector, and a tradition-bound and stagnant farm sector.

But in addition to such apparently intractable problems, special difficulties are encountered in plural societies—whether or not whites hold any significant political or economic influence. Assuming political representation determined by democratic process, the race having superior numbers can expect a legislative body dominated by members of their race and always sympathetic towards their interests. On the other hand, political strength in such societies is not necessarily drawn from people holding ingredients, such as technical skills and capital, essential to the growth process. Such is the position in Malaysia, where control of the Executive and ownership of the farm sector lie in the hands of the Malays: savings and the know-how necessary for the creation of a modern sector reside very largely with the Chinese. In view of the political bias towards the interests of Malays, it is difficult to see how the Malaysian government, in the shorter run, can encourage the Chinese to make their fullest possible contribution to economic development without offending the electorate.

In the Republic of South Africa, government of the plural society is not 'determined by democratic process'; nor can the Republic be classified as an under-developed country, except with respect to very labour-intensive (white and non-white) agriculture. Nevertheless, the industrialized and conspicuously affluent white communities remain scattered like enclaves in an economy where low income, lack of skills and other characteristics of under-development predominate. It would therefore be more realistic to regard South Africa as two economies, one industrialized and affluent, the other under-developed and, lacking independent 'poles of growth', entirely dependent on the transmission of growth from the former. In fact, growth is transmitted from the economically advanced 'white economy' to its backward 'neighbour' in very much the same way as growth was transmitted from the metropolitan powers to their former colonies during the nineteenth century.

But these two economies are not separate, certainly not in any real sense. The Bantu, having no alternative and little else to offer, move their labour across the 'frontiers' into the white farmlands or, in response to the occasional lowering of job reservation barriers, into urban areas. But the advance of the economy is rather like a goods-train shunting operation, with the white administration operating the 'engine' and 'backing-up' occasionally to take on an additional load of unskilled Bantu: the more 'trucks' of Bantu taken on, whether in the countryside or in the urban areas, the slower must be the journey to rising productivity. As to the prospects for separate development, it follows that the 'truck-loads' of Bantu in the rear can never hope to move into closer proximity to the 'engine': in fact, if Asiatics and Coloureds continue to be given priority, the prospect is that the Bantu will trail even further behind.

The system of job reservations is designed primarily to ensure adequate and 'appropriate' employment opportunities to the white population without danger of competition from equally skilled non-whites who might offer their services at lower rates of remuneration. But how can any government department, in a private-enterprise economy, hope to anticipate at all accurately the types and quantities of labour required in each and every industry, particularly when type of labour involves additional classification according to colour of skin? Rigid pursuit of such a policy must inevitably add a further element of uncertainty to private investment and serve to slow down the process of growth.

Products of this policy are readily apparent in recent experience. For example, a shortage of white bus crews in Johannesburg earlier this year led to the cancellation of so many services that on one day alone it was estimated that 15,000 white passengers were left without means of transport. Meanwhile the non-white transport services were operating so efficiently as to lead the Johannesburg *Sunday Express* to make the heretical suggestion that 'those standing in queues, waiting in vain for buses that had been cancelled . . . would welcome a bus to take them home—irrespective of the colour of the skin of the man at the wheel'. Similarly, the Chairman of the South African Railways Staff Association recently stated that, in the absence of an increase in the white labour force available or increased automation, the proportion of non-white labour used would have to increase markedly if the railways were to cope with the continuing growth in traffic.

In manufacturing industry, by far the principal 'growth centre' in recent years, manufacturers have been intent on making good their losses of white labour to non production-line jobs, as well as meeting their growing labour requirements, without regard to skin colour. Stressing the need for so-called 'influx control' but obviously angered by disregard for job reservation regulations, the government has now specified that no new industrial development involving the use of non-white labour will be permitted in twenty-three magisterial districts in the Transvaal, two in the Orange Free State and twelve in Cape Province without the personal authority of the Minister of Labour. Manufacturers were naturally quick to point out that such restrictions would serve only to retard economic growth. But even more embarrassing to the government, representatives of the white workforce, in whose interests such policies are allegedly pursued, proved equally antagonistic. In fact, J. A. Grobbelaar, General Secretary of the Trade Unions Council of South Africa, made it clear that he believes integration of white and non-white workers is necessary to any further industrial expansion.

The mere existence of limited political representation available to the Bantu in the conduct of their reserves and rigid off-the-job racial segregation are clearly insufficient conditions for separate development. In fact, economic integration has already reached a stage at which the whites are almost as completely dependent on the services of non-whites as the Bantu are dependent on the 'white economy'. In some manufacturing industries, whites now account for a quite small proportion of the workforce: for example, 32 per cent in chemicals, 16 per cent in food and only 11 per cent in both clothing and footwear.

How the government is likely to react to suggestions by white South Africans that even greater and more explicit economic integration should be permitted is difficult to anticipate. The ideological undercurrent in Parliament is still very strong indeed and it is extremely doubtful that the National Party will accept measures clearly in conflict with the theory of separate development if any tolerable alternative can be found. An increase in net immigration of whites would be politically the most expedient, particularly in view of the more rapid growth in non-white population. Alternatively, the government might seek to stimulate capital intensification in order to reduce relatively the need for less-skilled labour in manufacturing and other urban industries: but since the white farm sector is already too heavily encumbered by unskilled labour relative to its expected role in the development process, such measures could only result in increased unemployment amongst the Coloureds and Bantu and the risk of serious discontent. Meanwhile the level of investment and other prerequisites necessary to separate development in the Bantu reserves appear far beyond the foreseeable future resources available to either the government or the Bantu peoples. But perhaps external political pressures, rather than a need for reconciliation of the theory of apartheid with measures necessary to sustain economic growth, will exert a stronger influence on the future social and economic policies of the Republic.

## POLITICAL REACTIONS AND FUTURE PROSPECTS

In the U.N. Assembly on 10 October 1963 the representative of the South African government voted alone against 106 countries supporting a motion condemning racial discrimination in South Africa. Most Afro-Asian countries called for direct military intervention in support of world opinion. But the major Western powers, while denouncing apartheid, held that forceful intervention in the domestic affairs of a sovereign State would be illegal. Thus the trade sanctions ultimately imposed were the net product of militancy and caution.

In retrospect, it is remarkable that any government believed, or appeared to believe, that such sanctions would serve any useful purpose. Since these sanctions were never mandatory, some Western countries cut their exports and jeopardized considerable South African investments while others, notably Portugal, 'turned a blind eye' and capitalized on vacated opportunities.[8] Even if sanctions of a nature vital to the South African economy were made mandatory by the U.N., non-member nations would be free and equally willing to capitalize on such opportunities. Short of additional support by means of blockade, which almost certainly would lead to war, there seems little point in expecting sanctions to prove any more effective in the future than in the past.

In part due to traditionally heavy dependence on domestic sources of coal for heat and power, but also because alternative sources of petroleum and other essential raw materials were readily available, the South African economy has not yet demonstrated the hoped-for signs of weakness: in fact, with average growth in real Gross Domestic Product at a healthy 5·7 per cent per annum since 1962, the only immediate danger to the economy appears to lie in official reluctance to permit an increase in the intake of non-white workers on a scale necessary to sustain such growth.

[8] For Portugal, endeavouring to retain her colonial interests in Angola and Mozambique, there is more than 'a neighbourly interest' involved: prospects for her oil refinery at Lourenço Marques, particularly since similar sanctions were imposed against Southern Rhodesia, have never been brighter.

Nor has widespread foreign censure served to divide white opinion regarding the basic objectives of racial policy, despite recent criticism arising out of shortages of white labour in some sectors: on the contrary, the realization that they stand alone has only served to strengthen white nationalism to the point at which the Parliamentary Opposition now appears to accept that there can be no significant deviation from the course charted by Dr Malan in 1948.

With ways and means found to counter trade sanctions, a steady decline in foreign political pressures and apparently adequate means of ensuring internal security, it seemed that the Republic by 1966 was no longer under any compulsion to consider changes in her domestic policies. But many white South Africans have no wish to remain, in their own estimation, the 'political outcasts' of the world. Since early in 1967, and despite strong opposition from die-hard Afrikaners who threaten the first real break in National Party unity in twenty years, the Prime Minister (Mr B. J. Vorster) has pursued a much more outward-looking policy in an endeavour to improve South Africa's image abroad. In recent months, for example, the government has accepted the first black envoys to the Republic and reference has been made to the 'miracles' achieved by such 'rational men' as Jomo Kenyatta, Julius Nyerere and Milton Obote. Some have suggested that South Africa should contribute aid in various forms to the less-developed African economies. More recently, South Africa announced her intention to send a multi-racial team to the Olympic Games in Mexico City. But notwithstanding this effort to develop more cordial foreign relations, there are few outside the Republic—particularly in black Africa—who appear to believe that such concessions to foreign opinion foreshadow any significant change in racial policies at home.

Meanwhile, however, a new source of pressure has emerged with Southern Rhodesia's Unilateral Declaration of Independence and adoption of racial policies similar to those of South Africa. Although the black states earlier accepted, somewhat reluctantly, that the West refrained from the use of force against South Africa on the ground of legality, they can hardly be expected to accept similar reasons for refusal to intervene in Southern Rhodesia. Julius Nyerere, President of Tanzania and generally a moderating influence on black nationalism, has called on the principal Western powers 'to make clear whether they really believe in the principles they claim to espouse, or whether their policies are governed by consideration for the privileges of their "kith and kin" . . . if the west fails to bring down Smith, or having defeated him, fails to establish conditions which will lead to majority rule before independence, then Africa will have to take up the challenge' (Nyerere 1966). He admitted that the independent States of Africa, lacking sufficient military strength for such a task, would need allies; he also implied that, through default on the part of the Western powers, such allies would probably be sought within the Communist bloc. Thus these islands of white supremacy, rather than serve as bulwarks against Communism, might ultimately encourage alliances between African and Communist States.

Although primarily concerned, on this occasion, with Southern Rhodesia, Nyerere made it clear that Africans are now equally determined to put an end to white minority rule in the Republic and warned that 'there will . . . one day be learned the dreadful lesson that the whites constitute less than one-fifth of the South African population, and that numbers provide strength'.

Terrorist incursions from South West Africa, though met so far with relatively little strain on South Africa's security forces, are already a matter for some concern. Notwithstanding extremely heavy sentences imposed on terrorists taken prisoner, the recent resurgence of black nationalism suggests that an increase in the tempo of such activities can be expected in the future. Should these terrorists find it possible to forge a strong link with what the South African whites themselves regard as an inherent tendency towards violence amongst some of their own peoples, the situation could prove really explosive.

It is doubtful whether, in the event of serious disorder or open conflict, Western powers or the U.N. could do more than call for moderation and provide medical and other services designed to limit loss of life. But it is by no means certain that the black nationalists could force the South African government to come to terms. To weaken to the point of providing non-whites with equality of both political and economic opportunity would clearly put an end to their own existence. But is it likely that the black African States, if involved in open conflict, would subsequently settle for anything less?

The course of events over the past decade suggests that some form of compromise will ultimately prove necessary. Mr Vorster has already embarked on an outward-looking policy intended to improve South Africa's image abroad. But if that image is to change, the National Party must also modify its domestic policies, at least to the extent of removing the more obvious suggestions of racial inferiority, dismantling job reservation restrictions in occupations other than those directly associated with internal and external security, and providing direct (even if appreciably less than proportionate) Parliamentary representation to all races. Such measures alone would provide little immediate improvement in material welfare amongst the Coloureds and Bantu and would no doubt prove unpopular with many (although not necessarily a majority) of the white electors. But such changes are essential to any real sense of 'belonging' amongst the non-whites and in order to encourage these people to make their maximum possible contribution to economic growth. Nor would such a change of heart go unnoticed elsewhere: it would no doubt serve to temper future African demands and would certainly encourage overseas support for the white minority. In contrast, strict and unyielding adherence to Afrikaner ideology appears certain to lead South Africa towards ultimate collision with equally extremist black nationalism.

## ACKNOWLEDGEMENTS

The author is indebted to Professor E. A. Russell, University of Adelaide, for helpful comments offered on the initial draft of this paper. Errors of fact remaining and the views expressed are, of course, the responsibility of the author.

## BIBLIOGRAPHY

*Cape Times*, 30 April 1947.

Johannesburg *Sunday Express*, 21 January 1968.

MALAN, D. F. (1948). *Dr Malan's Policy for South Africa's Mixed Population.* Public Relations Office, South Africa House, London.

MANNING, A. W. (1964). 'South Africa and the World: In Defence of Apartheid', *Foreign Affairs,* **43**:1.

MASON, P. (1964). 'South Africa and the World: Some Maxims and Axioms', *Foreign Affairs,* **43**:1.

NYERERE, J. K. (1966). 'Rhodesia in the Context of Southern Africa', *Foreign Affairs,* **44**: 3.

PRICE, A. G. (1939). *White Settlers in the Tropics.* American Geographical Society Special Publication No. 23, New York.

Republic of South Africa, Bureau of Statistics: *Population Census, 6 September 1960.*

Republic of South Africa, *Statistical Year Book, 1966.* Bureau of Statistics, Pretoria.

G. ROSS COCHRANE

# *Problems of Vegetation Change in Western Viti Levu, Fiji*

*'Neither man nor his environment is static. Human effort ebbs and flows. Environmental factors appear and disappear.'*

A. GRENFELL PRICE *White Settlers in the Tropics*

## INTRODUCTION

ANY PARTICULAR AREA has numerous possibilities for use. Man has a greater power to modify his environment than any other living organism. Over the last 100 years the Fiji Islands, in the south-west Pacific Ocean, have had a rich and varied history of man-land interrelationships. Several different ethnic groups, each with strongly contrasted cultures, have adopted different methods in their use of land, and have placed different values on the resources of their environment. All have exerted important influences upon the landscape. Some of these have been localized; others have been more widespread and the tempo of change has increased over recent decades.

Fiji has a population of approximately 600,000. Average population density is between 75-80 persons per square mile, but this population is unevenly spread. Of the total 7,083 square miles of land in Fiji's 322 islands less than one-tenth is cultivated. Much of this cultivated land is found within the 1,900 square miles of the western half or 'dry zone' section of Fiji's major island, Viti Levu (Figure 17). Here population densities are high (from 600-1,000 persons per square mile of cultivated land), clash of interests is marked and conflict for land use acute. The problem of determining the most suitable land use—that likely to be of greatest benefit to the community as a *whole* in the *long run*—is a difficult one. The Fijian economy is based predominantly upon export of a vulnerable monoculture crop, sugar cane. Land ownership is almost entirely Fijian (84% is Fijian owned); sugar cane production is almost exclusively by Indians. Complex social and ethnic problems in Fiji's under-employed, multi-racial, plural society aggravate economic, political, administrative, and land tenure problems. Spate (1959) first drew attention to contemporary aspects of the

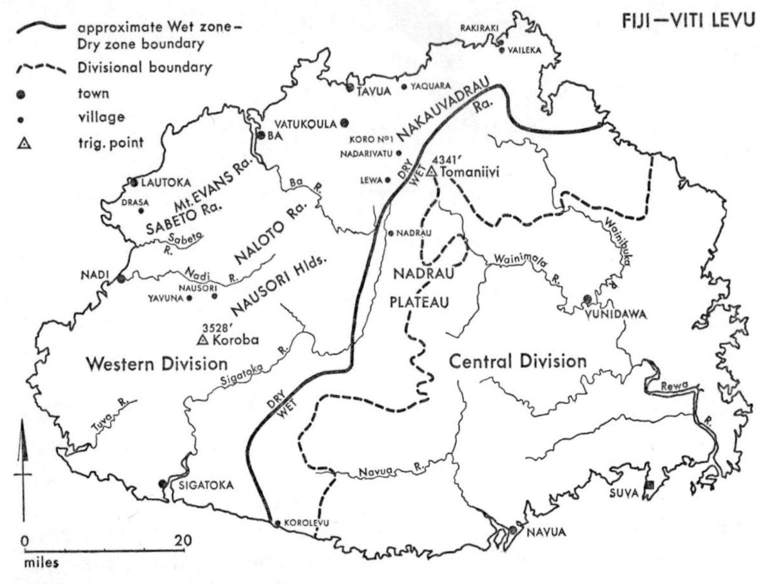


FIG. 17 Location map, Viti Levu, Fiji

colony's man-land relationships. This was followed by a Royal Commission investigation (Burns 1960) and detailed investigations by Frazer (1962) and Ward (1965). Current problems of land use in the dry zone of Viti Levu have been analysed recently by Cochrane (1967).

Population pressure upon available land resources is severe, especially in western Viti Levu. Fiji's population has more than doubled during the last twenty-five years: cultivated acreage has not shown a parallel trend. Rate of increase of Indians, who comprise over half the total population, is 4 per 1,000, and that of Fijians, who make up over two-fifths of the population, is 3 per 1,000. Both these rates are very high on world standards. They appear likely to continue. Frazer (1961) warned that Fiji's explosive population increase was outstripping the colony's ability and opportunity to expand the production of staple exports on which its prosperity depends. This population pressure upon limited land resources has contributed to the past and present misuse of land and environment in Fiji. This feature is not confined to this colony. It is all too common in young and developing countries. Recognizing the seriousness of this problem the current

Fiji Development Plan, 1966-70, has emphasized the need for diversification and a more rational assessment of resources (Council Paper 11/1966; C. P. 16/1966).

## PHYSICAL SETTING

Most of the Fiji islands, situated between 177 degrees east and 178 degrees west longitude and 15 degrees and 22 degrees south latitude, are steep, rugged, deeply dissected, volcanic islands. Viti Levu, the largest and major economic island, has many steep mountain ranges, up to 4,000 feet high. They are forest clad in the eastern half ('wet zone') where moist trade wind rains keep this windward side of the island moist throughout the year (Figure 18). Annual average rainfall is 120-140 inches over much of the area but rises to over 200 inches in the highest ranges. West of the central axial mountain ranges total precipitation is less. It ranges from 50 inches near the west coast to over 80 inches further inland. Orographic effects are marked so rainfall totals of 120-150 inches occur at elevations above 1,800 feet. There is a definite dry season in this leeward 'dry zone'. It varies from several months in a narrow coastal belt to one to three months inland at higher elevations near the physiographic and climatic divide. Cloudiness and relative humidity are less in the dry zone than in the wet zone. More detailed physical descriptions can be seen in Derrick (1957) and Ward (1965).

Twyford and Wright (1966) have shown that there is a very limited amount of productive agricultural soil in western Viti Levu, almost all of which has been developed already. Apart from alluvial lowlands, which are small, disconnected, largely coastal and intensively cultivated for cane growing, much of the dry zone is moderately steep to steep, dissected country supporting savannah-type grasslands, shrublands and woodlands (Figure 18, Plate 1). Some areas of bush, which formerly covered most of the area, remain. One such area, on the Nausori Highlands, is currently being milled for its indigenous *kauri*, podocarps, and various hardwood timbers (Plate 2).

## SETTLEMENT AND CHANGING FRONTIERS

The Fijian people are spread fairly evenly throughout the two main islands, Viti Levu and Vanua Levu (Ward 1959). Increasing opportunities are attracting villagers to urban centres, especially the rapidly expanding capital, Suva. Some changes are occurring

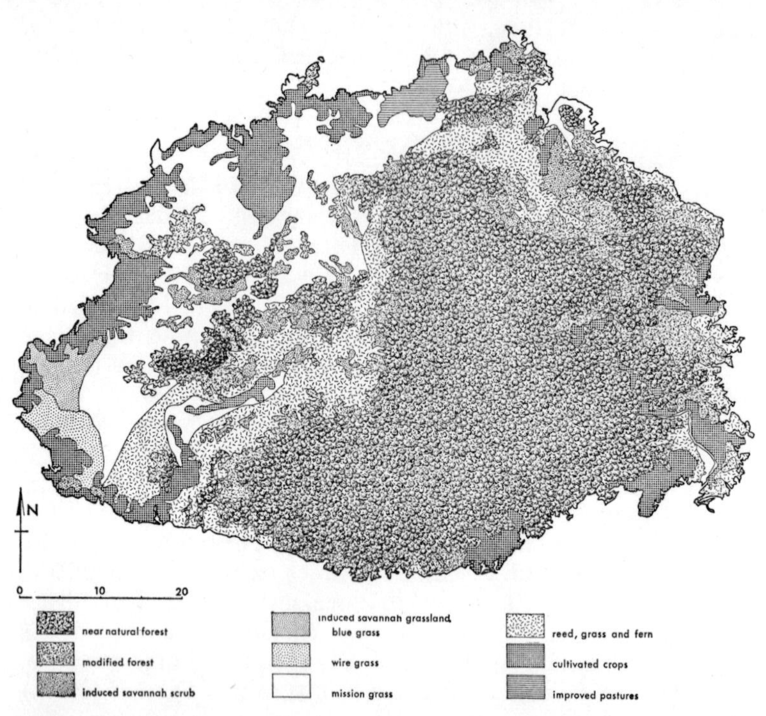


FIG. 18 Vegetation map, Viti Levu, Fiji

in Fijian agriculture (Frazer 1964), but a high percentage of Fijians still live in small villages (Ward 1960), much the same as they did in pre-European times. Fijians are strongly community minded and traditionally, a Fijian's first duty is to his community. The Fijian social system imposes many restrictions on an individual living or working where he desires. Land ownership is not by the individual but by the *mataqali*, an extended family group with all members having equal rights. Traditional customs, such as *kerekere*, where an individual is bound to share his belongings with his kinsmen, militate against individual enterprise. Similarly, traditions of hospitality and generosity, which make the Fijians very pleasant people, are not conducive to the accumulation of individual capital. Thus it is not surprising that two-thirds of the Fijian working males are engaged in village agriculture growing crops of *dalo* (taro), cassava, yams, *kumala* and bananas and papaya. Staple foods are augmented with fish and fruits and other food collected from the bush. An important cash crop is *yagona* (*Piper methysticum*), which is used in the preparation of the traditional drink, *kava*. Others are coconuts and bananas.

In village agriculture small areas are cleared out of the bush and sown to various crops (Plate 3). As production declines the gardens are abandoned and new areas cleared. The abandoned plots are invaded by the fringing bush or scrub species and regeneration occurs. This practice results in small local changes in vegetation; the environment is but little modified at any one time. However, evidence of former *dalo* terraces in present bush and savannah areas, quite distant from existing villages, shows that over a long period of time wide areas have been utilized in this shifting agricultural practice.

Traditional building materials, such as an indigenous reed, *gasau* (*Miscanthus floridulus*), were collected for thatch, for matting, for fish traps and for garden stakes. Timber for house frames was also gathered from the adjacent forests until recently. Although still used, especially in the more traditional villages, modern more durable building materials of sawn timber, weatherboards and fibrolite sheeting and corrugated iron roofing are becoming more frequent (Plate 3). More than 100 years ago Seemann observed that evidence of man's contact was almost universally present within the forests (Seemann 1862). However, vegetation changes were small and localized. The cycle of regeneration from village agricultural plots back to forest was relatively rapid. Fijians, prior to European contact, modified their environment to their own advantage so that these advantages were maintained without spoiling its power of productivity and usefulness.

Several phases of European settlement have each in their turn resulted in important and increasingly widespread modification of the environment. The first, in the early nineteenth century, was sporadic, temporary, and based upon the selfish, ruthless, commercial exploitation of the Fiji sandalwood, *yasi* (*Santalum yasi*) in the forests of Vanua Levu. The wood was highly prized in China for religious rites. The European traders' first concern with the forests was to gain large profits as quickly as possible. No thought was given to the preservation of the resources. Murders, cruelty and bloody massacres marred the short history of the Pacific sandalwood trade. The relatively small reserves of sandalwood trees were soon exhausted.

Later in the nineteenth century large quantities of firewood were cut around the coasts for processing bêche-de-mer—an item much sought after in the Orient as an aphrodisiac. Although the firewood was cut and collected by Fijians this demand for firewood was initiated by European traders. It soon led to a rapid depletion of forest cover from near the coast in parts of western Viti Levu. This was particularly the case in the narrow belt flanking the coast where rainfall was lowest and the dry period longest. Early European reports had noted the drier appearance and less luxuriant forest cover of this west coast area. The widespread disturbance of an ecosystem already in a delicate state of balance made conditions difficult for tree regeneration.

After 1874 when Fiji was ceded to Queen Victoria, large-scale European settlement began. Increased demands for firewood led to a rapid depletion of forest cover from near the coast in parts of western Viti Levu. This extensive clearing coupled with the later introduction of domestic livestock hastened the transformation of light forest vegetation to reed, scrub and grass forms. The development of sugar cane plantations by Europeans on the alluvial lowlands from Sigatoka in the south to Rakiraki in the north also contributed to the rapid change. The demands of the gold mining enterprise at Vatukoula, in the Tavua basin, for firewood, mine timber and land for crops also resulted in rapid inroads being made into surrounding light forest and forest areas.

The alien European culture introduced many new ideas and new techniques. It placed new and different values upon the island's resources. Prior to major European settlement, *dovu* or sugar cane was grown in small amounts by Fijian villagers. Some wild varieties are present in remote areas. Seemann first recorded sugar cane (*Saccharum officinarum*) in Fiji in 1861 (Seemann 1865-73). Europeans introduced the large-scale, intensive, commercial cultivation of sugar cane in Fiji in the early 1870s. Although land sales by Fijians were prohibited after 1874, Europeans soon leased the most productive alluvial lowlands from the Fijian owners. These areas were cleared, ploughed and sown into sugar cane. The Colonial Sugar Refining Company built their first mill in Fiji in 1880. The company, now called the South Pacific Sugar Mills Ltd., has three Viti Levu mills at Lautoka, Ba, and Penang. The fourth Fijian sugar mill is at Labasa, on Vanua Levu, Fiji's second largest island.

European commercial agriculture developed rapidly. The altered environmental conditions engendered by European enterprise resulted in a dramatic change in the vegetation. Within two decades sugar cane replaced former vegetation and village agriculture over an area of 100,000 acres. Today cane is grown in 130,000 acres in western Viti Levu (Cochrane 1967).

White settlement in the Fiji Islands has closely followed patterns demonstrated elsewhere in the tropics by Price (1939). White settlers have showed little tendency to settle permanently or to engage in physical labour. A high percentage of Europeans, even today, are expatriates—Price's 'sojourners'. This is particularly the case in administration and management.

The white settlers were loath to perform physical tasks associated with sugar cane growing, harvesting and processing. The indigenous Fijian people proved unsatisfactory because of attitudes, traditions and customs. Thus large numbers of indentured labourers were brought from India to work on the sugar cane plantations. These people introduced yet another set of traditions, attitudes, customs, and techniques to the island colony.

Price (1939) has shown that coloured people, if they are prepared to work harder and to accept lower standards of living, will drive out of tropic areas white groups which demand easier conditions of life. The Fijian Indians have adopted a very low standard of living, they have worked harder and have had much larger families than both the white and the Fijian peoples. Supremacy of whites and Fijians has been maintained by the creation of social barriers, laws, and political supremacy. The lack of realism and the injustices inherent in such a situation have been recognized by many (Spate 1959; Burns 1960). A recent legislative change, in 1967, although a step in the right direction, is merely a palliative. There is no simple answer: the problem is a complex one. The fundamental truths recognized by Price three decades ago hold equally as true in the jet age as they did in the age of discovery, four centuries ago. The delicate Fijian situation is fundamentally insecure (Frazer 1961). It is aggravated by problems of non-ownership and a lack of long-term responsibility inherent in non-permanent white settlers. The situation is doubly difficult for the segment of permanent white settlers.

The introduction of Indian labourers to the sugar cane fields quickened the tempo of change to the environment that began with the different attitude of Europeans to the island's resources. The Indians brought traditional peasant techniques of cultivation. They were uneducated, conservative, and accustomed to existing under an extremely low standard of living. Those that chose to remain in Fiji, by dint of hard work, gradually improved their lot. Many became tenant farmers leasing cane land and supplying cane to the mills. The traditional beasts of burden were the oxen which needed pasturage. Because of the shortage of good ground for cane growing none could be spared for supplying grazing for livestock. Although oxen do graze among the cane fields, at certain times this is impossible. On such occasions the stock had only the reed, fern, light forest or forest areas for possible browsing.

The Indian cane farmers, with no thought given to the long-term consequences, soon found that firing allowed stock better access to the bush. In addition, firing induced new shoots on some plants. These were palatable. Uncontrolled, irresponsible and illegal firing became a common and accepted practice. Fire, one of the most potent ecological factors, in the hands of Indian cane growers brought about a modification of the vegetation of western Viti Levu that was unrivalled by any other in its long history of man-land relationships.

## FIRE AND VEGETATION CHANGE

Today, communities of grass, scrub and open woodland of light forest cover most of dry zone Viti Levu (Figure 18, Plates 1 and 4). Formerly forest covered most of this area (Seemann 1862; Parham 1964). The present savannah-type vegetation owes its origin to man-set fires over the last ninety years (Parham 1955). It is artificially maintained by the frequent firing that has been practised widely over the last fifty years. There is scant evidence that grasslands were ever a significant part of the natural vegetation complex of dry zone Viti Levu. There is a relative paucity of indigenous grass species in the Fijian flora. Many of the few indigenous grasses are wet zone grasses confined to wet habitats. Seemann (1862) recorded less than a score of grasses in Fiji in 1861. Several of these were exotic species brought by early European settlers. Grasses, especially savannah-type grasses, are not an important element in the indigenous plant cover of Viti Levu (Parham 1958).

Fire has always been present in western Viti Levu. Occasional though very rare lightning fires have occurred. These would rarely burn far because suitable fuel to carry fire was largely lacking under natural conditions. Most lightning in this part of Fiji is accompanied by torrential rain so fires would be unlikely. If begun before rain, normally they would be extinguished by rain shortly after.

Fijians have doubtless also used fire to prepare agricultural plots. Such fires would have been small, controlled and of very localized occurrence. If fires sometimes escaped into fringing bush, the vegetation would be able to regenerate through stages of *gasau* and secondary bush to bush. It would not have been subjected to repeated fire at frequent and regular intervals.

Continued, frequent burning as practised by the Indian cane farmers, and to a lesser extent over recent decades by Fijians also, has led to a retreat of bush margins. Fire-induced, exotic grasslands have extended rapidly. Main species are mission grass (*Pennisetum polystachyon*), wire grass (*Sporobolus indicus*) and blue grasses (*Dicanthium* spp.). Nadi blue grass (*D. caricosum*) was introduced to Fiji in 1907. It is widespread in the south-west of Viti Levu, where it is an important pasture grass (Figure 18). The other two members of this genus in Fiji, Vuda blue grass (*D. annulatum*) and blue grass (*D. aristatum*) have a similar distribution, but are less important as pasture grasses. The C.S.R. cattle station at Yaqara is fostering the growth of these grasses. Mission grass, an African elephant grass, was introduced to Fiji via a United States agricultural station, in 1920. It has proved utterly worthless as a pasture grass. It has spread widely following firing and now covers extensive areas in western Viti Levu (Figure 18, Plate 1). Wire grass is less aggressive and less widespread than mission grass. However, it is also an unpalatable, worthless grass. Like mission grass its spread and maintenance are favoured by frequent firing.

Areas of savannah-like scrub, often dominated by the exotic guava (*Psidium guajava*) have spread so vigorously in the wake of frequent fires that this species has been declared a noxious weed (Parham 1958). In areas that have been frequently burned, then kept free from firing, another exotic scrub tree, the legume *vaivai* (*Leucaena leucocephala*) forms important communities. *Wase* (*Dodonaea viscosa*) is a common indigenous shrub in areas relatively free from firing.

Repeated firing has resulted in many depauperate areas of bracken (*Pteridium esculentum*) and *Dicranopteris linearis*, developing on land that became badly worn out and eroded through repeated firing. Formerly such communities were chiefly associated with lateritic *talasiga* or 'sunburnt' soils. Repeated firing on *talasiga* areas has resulted in severe erosion and the complete denudation of large areas. Vegetation disappeared from the Ba Closed Area, a large undulating to hilly area south of the Ba alluvial lowlands, as a result of repeated firing.

Some fire-tolerant native species, like *nokonoko* (*Casuarina equisetifolia*) have become more common than formerly as a result of firing. These sparse communities are very different from the former forest (Figure 18).

The extent and increasing frequency of illegal fires has caused increase in surface runoff. Runoff coefficients from sparsely grassed catchments are as high as 0·90. They are 0·45 from catchments with some bush, scrub and grass cover, and 0·15 and less from undisturbed bush catchments. The Fijian Departments of Forestry and of Agriculture have been concerned over the increasing erosion resulting from widespread indiscriminate firing in dry zone Viti Levu (C. P. 1950 *et seq.*). The following examples of fire-induced vegetation change in western Viti Levu embrace a wide variation in location, climatic, lithic, topographic and edaphic conditions. They demonstrate the role that indirect effects of sugar cane farming have produced. The vegetation at Lewa, Nadarivatu, Koro No. 1, Yavuna, Nausori Highlands and Drasa after frequent firing was predominantly mission grass (Plates 1, 5). In much of the Ba Closed Area bare eroded slopes devoid of any vegetation were the legacy of repeated firing (Plate 6). Ruthless disregard of the environment's resources characterize this phase of man-land relationships in western Viti Levu. The Indian cane farmers modified the non-cane areas to provide grazing for cattle. Their methods led to a deterioration of the productive capacity of the environment. This was a marked change from the pre-European Fijian's husbanding of environmental resources.

## VEGETATION CHANGE IN THE NAMOSI CATCHMENT

Detailed comparisons made between Yavuna and adjacent flanking steeply sloping areas of Upper Pliocene basaltic agglomerates to the north and of Miocene andesites to the south demonstrate that active natural erosion is acutely aggravated by contemporary anthro-pyrogenic practices.

The Yavuna area is an elongated, maturely dissected, hornblende-granite mass with rejuvenated downcutting resulting from Quaternary uplift. The granite has been intruded with a multitude of basic andesite dykes (Bartholomew 1960). The mass strikes north-west, extending for seven and a half miles, from near Yavuna to two miles beyond Tubenasolo (Figure 19). It is only

PLATE 1 Yavuna area. Firing has destroyed most of the bush cover of this area. Fire-induced and fire-maintained exotic vegetation of mission grass, guava scrub and bamboo thickets covers most of this deeply dissected, granite hill country.

PLATE 2 Bush, Nausori Highlands, western Viti Levu, composed of nearly pure stands of *yaka*. Large areas of such forest were earlier destroyed by indiscriminate firing.

PLATE 3 Small plot of Fijian agriculture cleared out of the forest. Crops include banana, *dalo*, *kumala* and cassava.

PLATE 4 Tavua basin, western Viti Levu. Savannah-scrub vegetation in the foreground and the extensive mission grass areas beyond are the result of reckless firing. Experimental planting of exotic conifers, *Pinus elliotii*, from the Nadarivatu forest station, can be seen beyond the scrub.

PLATE 5 Yavuna area, western Viti Levu, showing prevalence of slips on recently burned areas.

PLATE 6 Ba Closed Area, a fire-induced desert. Slow colonization of flats by mission grass has begun after several years of closure. The background slopes are still barren and devoid of vegetation.

PLATE 7 Ba basin from the north. Note the eroded ranges of the Ba Closed Area in the background and the sparse cover of grass and guava in the foreground. Sugar cane farming has spread from the lowlands into many submarginal upland localities. Firing has destroyed former bush.

PLATE 8 This *talasiga* fern, *qato*, is all that can grow on this formerly productive lowland of the Ba Closed Area. Repeated, indiscriminate, and illegal burning of this area produced an eroded desert.

K

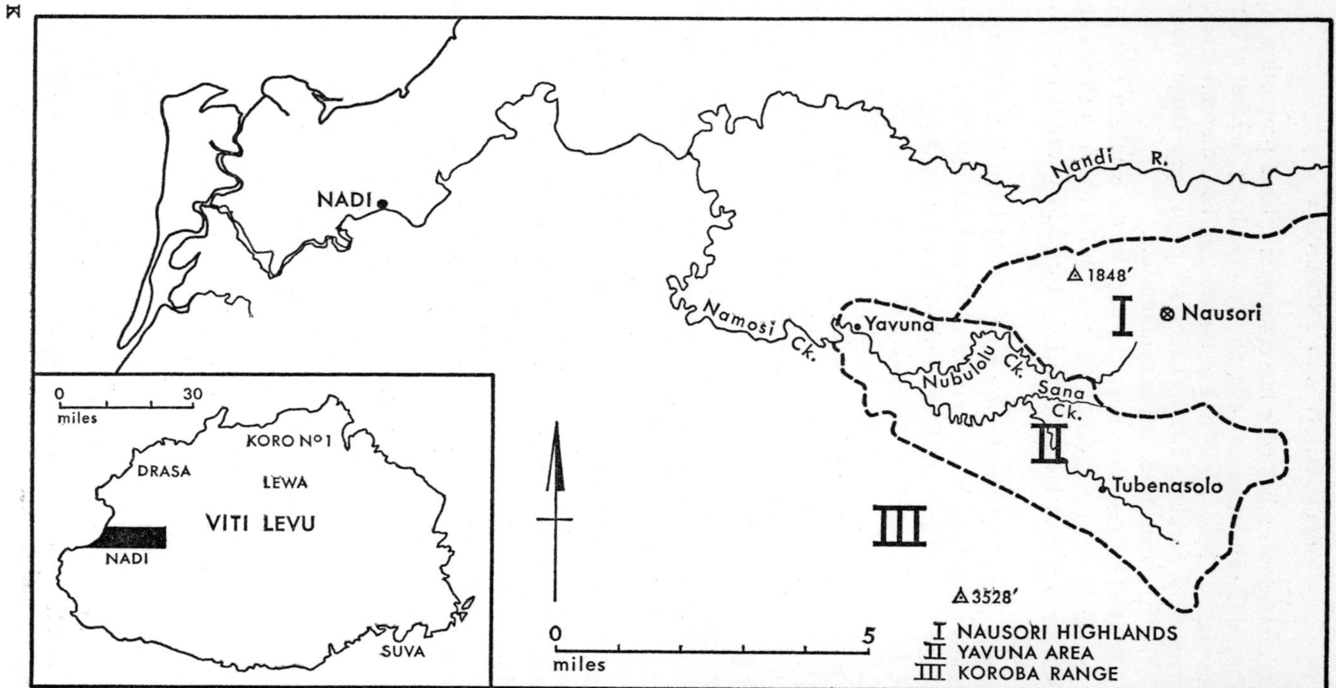


FIG. 19 Location map of Nausori Highlands, Yavuna, and Koroba Range, Western Viti Levu, Fiji

approximately two miles wide but comprises a large proportion of the Namosi Creek catchment. Elevation ranges from 200 feet at the eastern end to above 1,500 feet at the western end. Summit levels are commonly between 200-400 feet although they range from 100-700 feet. Drainage texture is medium-fine.

It is flanked on the north-west and west by the southern escarpment of the Nausori Highlands. On the south-east steep slopes of andesite flows rise to a maximum height of 3,528 feet at Koroba (Figure 20).

Yavuna is well within the dry zone of Viti Levu but, although only eight miles from Nadi (rainfall *c.* 75 inches) rainfall is probably near 90-100 inches. This results from the flanking position of the highlands and their funnelling effect on clouds. Rainfall is markedly seasonal, falling mostly in the form of short-duration, high-intensity convective storms during the wet season (November-April). Despite the marked maximum of rainfall during the wet season sufficient rain falls during the dry season to keep soils under forest moist throughout the year.

Formerly, forest covered the Yavuna area. Sufficient common species occur in the secondary bush remnants to indicate continuity of forest cover (albeit a lighter tropical forest on the lower slopes) between Nausori Highlands and Koroba Range (Plate 2). Today, as a direct result of firing and livestock grazing, much of the area supports a sparse cover of introduced mission grass, guava bushes, and indigenous ferns, *qato*, or *Dicranopteris linearis* (*talasiga* fern) and bracken; *Miscanthus floridulus* (reed) flanks burned bush. Small areas of *nokonoko* are also present. Extensive thickets of exotic bamboo (*Bambusa vulgaris*) extend up numerous valleys.

This vegetation, unlike the bush areas, does not bind the Vunatoto gritty clay loams, Yavuna sandy loams and the Vatubabu stony clays of the Yavuna area (Twyford and Wright 1966). Evaporation is also high and the soils are commonly dry for long periods. Consequently, the frequently fired and grazed areas are characterized by a depauperate vegetation of grasses, shrubs and ferns, thin soil cover and severe stripping of soil. Widespread exposure of weathering parent material of pinkish-white granite and reddish-brown andesite dykes is present.

Vigorous sheetwash is widespread and slips are common. Corrasion (e.g. pot-holing) occurs along stock tracks and gullying is present in many of the older eroded areas. Infiltration capacity of the thin soils is minimal, thus surface runoff is excessively high.

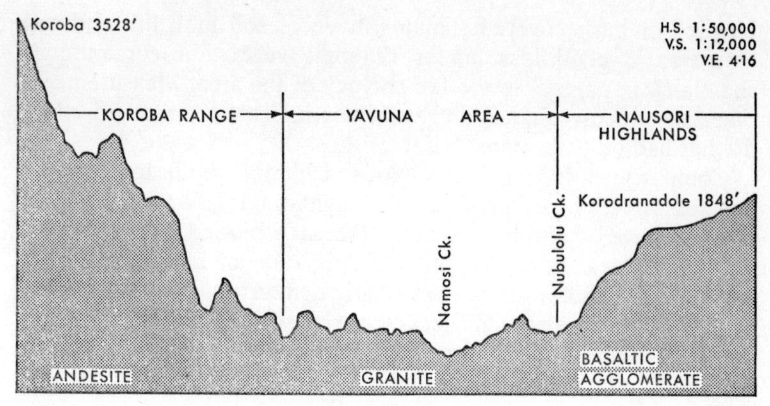


FIG. 20 Cross-section from Koroba through Yavuna area to Nausori Highlands, Fiji

The continuous downslope movement of materials, the rapid rise of streams following rains, and the steady deterioration of the area constitute a concomitant siltation-flooding problem downstream on the Nadi lowland cane areas. This flooding hazard is augmented by the rapid runoff, though less severe sheetwash, from the steep, flanking, grassy, highland slopes where the bush cover has also been destroyed by firing.

Broad general comparisons made of soil type (depth and structure), lithology and vegetation cover (species and density) between the Yavuna area and the flanking slopes clearly reflect the influence of firing.

Detailed studies were also made within the Yavuna area. Measurements were taken of angles of slopes, the incidence of slips, the type of vegetation, the recency of firing, and the presence of sheetwash. The position of occurrence of slips was noted and their dimensions recorded. The relative movement of materials by surface drainage lines, by sheetwash, by gravity, by saltation through raindrop impact, by gullying, and by soil creep, was noted for slopes, new and old slips, and in valleys in the grass (burned) areas. These were compared with soil movement processes under bush-covered areas within the survey area.

Measurements of soil depth, particle size, soil structure, soil biological structures, humus and vegetational debris, soil surface condition, amount, and kind of vegetational cover made in comparable valleys were analysed to assess the damage caused by burning of bush cover. Estimates made of soil moisture and soil temperature conditions under different vegetation communities show a close parallel to the fire ecology of the area. Measurements taken of stream volumes and velocities indicate the marked effect fire has had on catchment characteristics.

Comparisons in slope conditions, including both longitudinal and transverse valley profiles and longitudinal valley side profiles in forest-covered and burned areas (recently burned, several burns, long burned etc.), carried out for estimation of infiltration capacity/surface runoff ratio of soils clearly demonstrate the role played by fire. The prevalence of occurrence of slips in the burn-grazing sequence from bush to bare land shows a direct relationship to frequency and recency of burning.

*Nausori Highlands Escarpment and South-western Rim of Namosi Creek Catchment*

Both the Yavuna area and the Nausori Highlands escarpment have had most of the bush destroyed by firing. On the latter, the Keiyasi stony and bouldery clays (nigrescent soils), and to the east, the Nadarivatu clays (steepland humic latosols) of the highest slopes appear much more stable than the red-yellow podzolic soils of Yavuna. In addition, these former soils support a more vigorous grass-shrub-low tree vegetation on burned areas than is the case on the Yavuna area.

On the escarpment slopes, pyric induced communities of *Pennisetum polystachyon* with some *Sporobolus indicus* (wire grass) are widespread. Canopy cover of grass vegetation (except where burned within the last season) is near 100 per cent but basal cover is little over 50 per cent. Numerous herbs are present, notably, *Crotalaria striata, Polygala paniculata,* blue rat's tail (*Stachytaphetta urticaefolia*), *Desmodium heterophyllum, Sida* spp., peanut 'grass' (*Atylosia scarabaeoides*), and tar weed (*Cuphea carthagenesis*). Low, shrubby, hibiscus burr (*Urena lobata*), a declared noxious weed, and shrubby Chinese burr (*Triumfetta rhomboidea*) are also common. However, even where these plants and grasses are well established, basal coverage is rarely more than two-thirds. Consequently, shrub and tree communities are more localized, particularly flanking stream courses and seepage areas. Guava is common as shrub thickets. Shrubs and low shrubby trees of *nuqanuqa* (*Decaspermum fruticosum*), *Wickstroemia indica, drou* (*Parasponia andersonii*) syn., *wase* and *vaivai* also occur. They are small and scattered, as all the slopes, other than bush areas, had been burned at least once, often more frequently, during the three years prior to 1966, when field investigations began in this area.

Erosion by raindrop impact is present on recently burned areas with bare earth. This is particularly important in removing fine particles in burned areas of *gasau* where many angular unweathered rock fragments litter the bare soil.

*South-eastern Flanks*

Similarly, beyond the southern limits of the Yavuna hornblende-granite mass, the Vatubabu stony clay (red-yellow podzolic) developed on andesites appeared more stable than similar soil derived from the granite. However, although the slopes appear well covered with grass, closer inspection shows considerable bare earth between the tussock clumps, even under vegetation unburned for several seasons. Sheetwash is present, though not critical, but surface runoff is high. In view of the volume of water moving as rapid surface runoff and its resulting fluctuations in stream flow it is distressing to observe unnecessary and indiscriminate firing continually pushing the bush line upslope to the mountain crest. The grazing potential of such steep slopes and stony soils is extremely limited. Furthermore, the firing of bush and the subsequent pyric-induced grass growth is detrimental to the sound watershed management of the Namosi Creek catchment.

*Yavuna Area*

Despite the division of this area into numerous segments by pockets of bush and bamboo, fire frequently burns for a day and sometimes two days. Fires impinge on bush margins gradually reducing them in area.

Investigations show that most burned-over *Pennisetum*-clothed slopes in excess of 20° have active slips. Angles of slopes where slips are most common are generally in excess of 25° (25-35°). Valley head slopes are very steep, often 50°, but slips are noticeably absent. Figure 21 shows representative slopes and prevalence of slips in the Yavuna area.

Most fired slopes examined showed evidence that over seventy per cent of the surface had slip scars of different ages. Many of these appear, on superficial examination, to be stabilized by later growth of vegetation. Closer scrutiny demonstrates that though some plants are growing tenaciously on old slip faces, coverage is minimal and sheetwash keeps pace with weathering so that soil is not developing. Most slips appear remarkably static. Observations from Fijian villagers support this belief. Upslope movement of slips is negligible. The appearance of upslope fretting and progressive grassing below recorded by Marsh (1965) results from the partial covering of grass downslope by sheetwash materials. Following burning of bush, slips commonly occur near the crest at the break of slope between convexity and straight slope where bush formerly reached the crest. Slips are prevalent at the flanks of recently burned bush regardless of slope conditions.

Dense growth of *Miscanthus floridulus* commonly follows the first firing of bush but a second fire greatly reduces reed growth and hastens soil depletion. Reeds rarely survive two firings, being replaced by scrub and grass growth. In places, bamboo thickets extend along the valley floors following the firing of bush.

The position of slips follows no particular pattern other than that shown in Figure 21, occurring randomly on slopes of 25-35°. However, the size of slips shows a tendency to uniformity. These are normally sub-circular with a diameter of 25-30 feet and rarely more than 18 inches deep. Some slips are much larger, and some are even linear, but the saucer-shaped slips predominate markedly. Some old slips, deepened by gullying, have back walls several feet deep. On one severely eroded hillside numerous small slips had vertical back-walls 3-4 feet high. This represents a more advanced stage of erosion than that of non-continuous slips.

*Firing and Sheetwash*

More important than the obvious erosion by slips is the removal of soil by sheetwash following firing. Vigorous removal by sheetwash is active on both bare parent material and on the thin soils of grassed slopes. Once the slopes have been burned of vegetation through firing or, at best, are clothed with a thin cover of grass (a direct result of the anthropic factor), the surface layers of whole slopes can actively move downslope during heavy rains. Rapid weathering into coarse grits from granite, or into fine silts from andesite dykes, the movement of both channelled and free water (rivulets and sheetwash), gravity movement (colluvium) and saltation by raindrop impact ensure continued rapid movement of materials downslope; this is only partially retarded by the sparse covering of grass, fern, and occasional shrubs.

The relative importance of mass movement processes differs greatly between forested areas and burned areas within the Yavuna

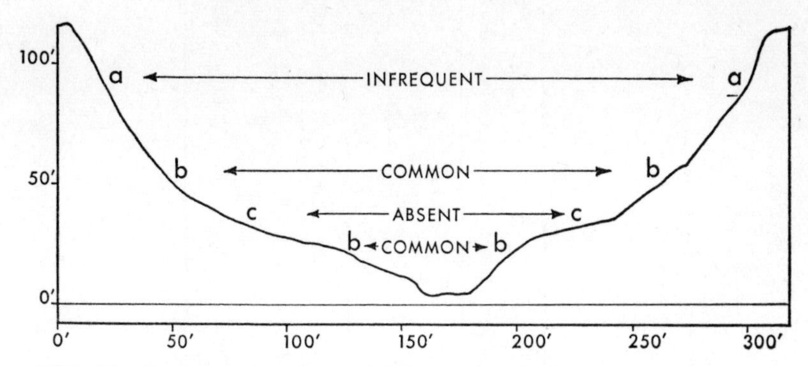


FIG. 21 Representative slopes and frequency of slips, Yavuna, Fiji:

| *Slope categories* | *Frequency of slips on fired slopes* |
|---|---|
| a 50° | Infrequent |
| b 25°-35° | Common |
| c 12°-15° | Absent |
| d 25°-35° | Common |

and flanking areas. Within relatively undisturbed bush deep, moist soils with organic accumulation tightly bound by a maze of roots from trees, shrubs, herbs, and fern rhizomes show little evidence of surface movement. The several canopies of vegetation plus surface organic litter on the soil are an effective safeguard against raindrop erosion and vigorous sheetwash. Slips are extremely rare and gullying is absent. Most downslope movement appears to result from soil creep. Some slumping of highly saturated clays flowing out from under tree roots was observed in bush areas following heavy break-of-season rains. This was not common. Infiltration capacity through organic litter sponge and deeper soil is much greater than on adjacent grassy slopes with their thin, largely bare, chiefly mineral soil. Stream regimes dramatically demonstrate these differences (Table 1). Subsurface seepage is probably important in such forest areas.

On burned areas sheetwash undoubtedly is of greatest importance in mass-movement. Surface runoff from the thin soils is considerable. After heavy rain whole hillsides are mobile. In the Yavuna area coarse granite grits as large as $\frac{3}{8}$-$\frac{1}{2}$ inch in diameter as well as finer particles are moved downslope in quantity by sheetwash. Measurements of sheetwash from a sparsely vegetated, eroded area 141 square feet in extent before and after a two-hour rainstorm showed 4·5 cubic feet accumulated at a slight flattening

TABLE 1

*Stream Flow*

| Stream | Catchment Vegetation | Minimum Flow | | | Maximum Flow | | | Sustained Flow | | | |
|---|---|---|---|---|---|---|---|---|---|---|---|
| | | Date | Rain | Cusecs | Date | Rain | Cusecs | Time: | Cusecs | Cusecs | Rain |
| | | | | | | | | to peak in hours | at 12 hours later | at 24 hours later | |
| Nubulolu Creek | Largely mission grass | 18/1/66 | Nil | 12·85 | 19/1/66 | 4·5″ | 3,874 | 2 | 52·0 | 22·35 | Nil |
| Sanna Creek tributary | Bush and partly grassed (small) | 18/1/66 | Nil | 22·00 | 19/1/66 | 4·5″ | 528·3 | 5 | 72·6 | 66·72 | Nil |
| Tubeidreketi Creek | Dense bush | 15/1/66 | Trace | 440 | 16/1/66 | 2·2″ | 568 | 7 | | 582 | Nil |

at the base of the area. Evidence of greater amounts washed over the slope was obvious. The area was bare before the rain. Sheetwash from grassed slopes has greatly infilled the upper reaches of many valleys. Measurements on several valley floors show depths of over three feet. Infill possibly occurs to maximum depths of over 20 feet (see Figure 22). This is particularly significant as the tributary streams in burned areas do not demonstrate recent rejuvenation. This is masked by the contemporary filling of valley floors with sheetwash materials following firing. Incised meanders and river terraces are present on larger streams. Also rejuvenation can be traced up tributaries within bush catchments.

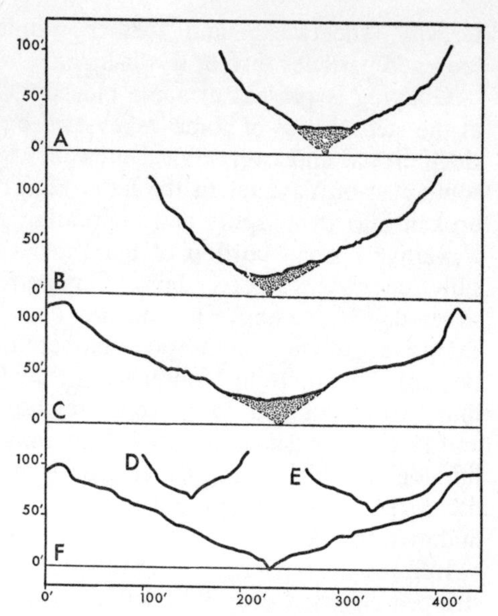


FIG. 22 Valley Profiles in the Yavuna area. A, B, and C, fired areas. D, E and F, forested areas

Vigorous corrasion occurs during high intensity rainfall, causing widening and deepening of channels. Ephemeral surface drainage, particularly along stock tracks, transports large quantities of weathered materials. Within a distance of less than 50 feet, one 9-inch-wide stock track along a ridge crest accumulated 1½ cubic feet of coarse, ⅜ inch, granite grit in an aggradation hollow following a severe two-hour rainfall. Stock track watercourse spillways at saddles on ridge crests often plunge down steep (50°) valley heads and not uncommonly gullying has developed. Such gullying appears to cause lateral slumping and slips.

On bare soil and bare weathering parent rock, saltation of particles by the physical impact of raindrops is a potent erosional factor. Examples were measured of granite grits of ⅜ inch and ½ inch diameter being knocked laterally distances of from 9 to 15 inches by individual raindrops. On steep slopes the lateral trajectory of these particles caused them to fall considerable distances.

Gravity, sheetwash, and further raindrop impact move these loosened particles further downslope.

Gullying is present at some old slips, along some stock tracks, at the steep heads of some valleys fed by temporary streams from ridge crests, and over a local area of tuva clays (*talasiga*) in the south-east of Yavuna. In the latter area ridge slopes have become broken into steep spurs and alternating gullies but this comprises a relatively small portion of the Yavuna mass. Granites are possibly deeply weathered here. Simonett (1967) observed both single debris avalanche movements on less weathered granite and extensive gullying on deeply weathered granite bedrocks in the Bewani and Torricelli Mountains of New Guinea. He also observed that there appeared to be some relation between number, type, and size of landslides in granites in contrast to those on different lithologies. This also appeared true in dry zone Viti Levu. These lithologic differences are enhanced by burning and vegetation denudation in Fiji.

Infiltration capacity is virtually nil on slip faces and severely stripped slopes. Conversely, surface runoff is nearly 100 per cent. The large number and close spacing of impermanent surface drainage lines demonstrates this. Small runoff channels scoured across the surface of one slip were at intervals of 2 inches to 4 inches and incised to depths of ½ to 3 inches. Infiltration capacity is little better on the thin soil of grassed areas. The shallow depth, coarse texture, loose structure, dryness, and limited amount of humus and vegetational debris of the gritty, fire-modified soils are conducive to a rapid wetting and extensive movement of particles by sheetwash.

Although the better grassed areas (largely ones where fire has been absent for two or more seasons) have an 80 to 100 per cent canopy coverage, basal coverage is often only 25 per cent. This is much less than on the flanking hill slopes of the Nausori Highlands. Raindrop erosion is not important but sheetwash is very active. Surface runoff is rapid and soils quickly dry out again. Soil biological structures (roots and rhizomes) have little ability to hold the thin soils under grass communities. Figure 23 shows a bisect exposed by a slip that occurred on 19 January 1966. Note the sparse ground cover despite a closed canopy cover from the three-year old vegetation. A sparse root 'network' is largely confined to the top five inches of dark red-black gritty clay loam. Only the rhizome of *Pteridium* extends to the coarse red grits of the C

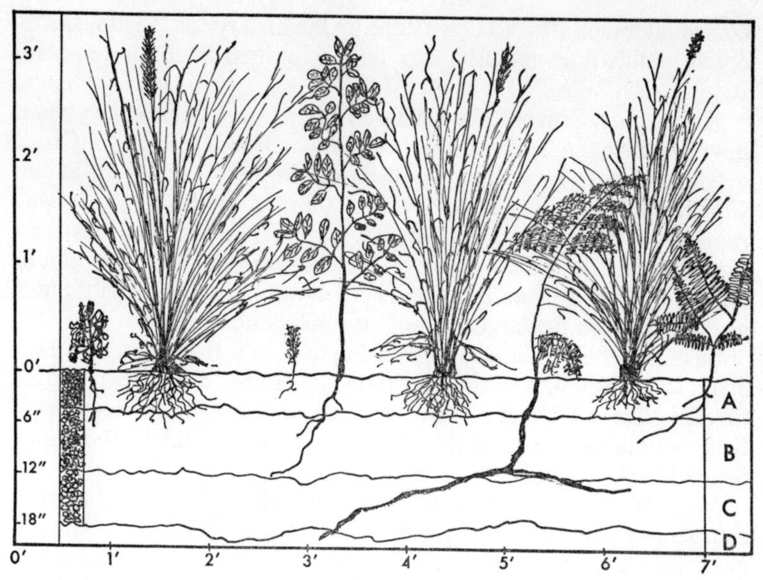


FIG. 23 Sparse root 'network' of grass communities, Fiji

horizon. The grass is *Pennisetum polystachyon*; other species from left to right are *Wickstroemia indica, Polygala paniculata, Psidium guajava, Pteridium esculentum*, and *Dicranopteris linearis*. This contrasts markedly with the webbed inter-penetration of forest plant roots. Soil temperature and soil moisture conditions under grassland fluctuate to a much greater degree than in the deeper soils under bush cover.

Sustained flow of clear water without great flood peaks followed by troughs characterized tributary streams of the Namosi Creek with forest catchments. Normal bankful discharge was not achieved during observations despite heavy rainfall. Silt load was not great. During the same period of field observations, tributaries draining adjacent grass-covered catchments showed an enormous range in flow, a 300-fold increase occurring within two hours from commencement of heavy rain. This had dropped to only a four-fold increase within twelve hours. Twenty-four hours after peak flow the stream flow was back to one and a half times minimum flow (Table 1). Normal bankful discharge was greatly exceeded. Vast quantities of coarse granite grits were disgorged into Namosi Creek. The turgid, turbid torrents carried great amounts of both fine and coarse materials washed downslope by the high runoff. Undoubtedly much of this finer material was later deposited in the lower Nadi River.

Although catchment sizes differed the relative figures for stream flow are revealing. The essentially grass-covered Nubulolu Creek catchment was larger than the bush catchment tributary of Sanna Creek. Nubulolu peak flow of 300 times minimum flow was reached in two hours. Peak flow in the other stream was less than one-sixth of this and was reached many hours later. Also minimum flow from the smaller bush catchment was one and a half times as great as from the larger Nubulolu catchment.

Sustained flow was also higher, being nearly three times greater than the Nubulolu, twenty-four hours after rains. Nadarivatu figures for Tubeidreketi Creek, with a bush-covered but larger catchment than Nubulolu Creek, demonstrated a similar trend to the small Sanna Creek tributary.

Measurements of transverse and longitudinal profiles of closely comparable tributary valleys of the Koukou and Sanna Creeks indicated that valley wall slopes, though steep in bush-covered pockets, were more regular than their fired, grass-covered counterparts. Slips and sheetwash were not obvious on bush slopes. On frequently burned areas valley walls were being steepened and becoming more irregular through slips. Surface area exposed to weathering and sheetwash was markedly increased as former uniform slopes were gradually being fretted to an irregular secondary pattern of spurs and valleys through slip erosion. The deep infilling of valley floors with sheetwash and colluvial materials that was such a feature of the burned areas was absent in the bush areas (Figure 22).

As observed by Marsh (1965) the erosion problem of the Yavuna area stems chiefly from fairly regular burning. Following heavy, break-of-season rains twenty new slips occurred on 19 January 1966 on an area of 600 acres, west of Tubenasolo, that had been fired only a few weeks earlier. Ground cover by vegetation was virtually non-existent. Erosion by slips and sheetwash was severe. In another small area two slips appeared on slopes burned only a few months previously. In contradistinction, within the detailed sample area only one slip occurred on grass slopes that had an undisturbed three-season vegetation growth (Figure 23). No slips occurred on dense grass (more than three years free from fires) nor in bush areas. There appears to be direct relationship between frequency and recency of firing, prevalence of slips, and severity of sheetwash erosion. Similar conclusions were reached by officers of the C.S.I.R.O. Division of Plant Ecology working in Queensland tropical areas (Stacey 1968: pers. comm.).

The Yavuna area and flanking highlands of the Namosi Creek catchment should be returned rapidly to some form of bush cover. Anti-burning publicity, which began in the 1950s, has had some effect in the Yavuna and adjacent areas. In areas free from fires young stands of *nokonoko* are present now that could not be seen on 1951 aerial photographs. Similarly areas of guava, *vaivai*, *wase*, and numerous other shrubs and low trees have prospered in the absence of fire. In some of the more remote areas that have largely escaped firing, scrub and low tree cover has increased from less than 15 per cent to around 25 per cent. The activities of a special fire ranger in the area have effectively reduced, though not stopped, burning since 1966. Regeneration from grass, through various exotic and indigenous shrubs, to low trees and some form of bush cover can be achieved if fires are kept out of this area. Scrub and bush cover has increased 100 per cent in areas totally free from fire over the last fifteen years (Marsh 1967: pers. comm.).

Ideal watershed management should give rise to maximum total runoff, that is well distributed throughout the year, and with a minimum of erosion. The relative economic values of forestry and of grazing for this area weigh heavily in favour of forestry. Pastoral potential on the poor soils is extremely low even allowing for heavy capital expenditure. As Marsh (1965) observed, at present the area is not even lightly grazed on a permanent basis, yet erosion is severe. Although some stock do periodically graze the area, the introduction of grazing stock on a more permanent basis would merely accelerate the rapid deterioration of the area.

In view of the success achieved by the planting of conifers on eroded soils at Lewa, Nadarivatu, Koro No. 1, and Drasa, the best land use for this whole area of the upper Namosi Creek catchment would be forestry. This would couple the best economic use of the land with sound ecological principles of conservation and catchment management.

Lack of finance, acute land tenurial problems, and difficulties of maintaining anti-burning policies should not militate against the execution of afforestation programmes. Diversification of Fiji's economy is an urgency (C. P. 11/1966, C. P. 16/1966). Forestry would be a form of land use that would maintain and increase the permanent productive capacity of this land resource. Present land use, using fire, is impairing the conservation of this resource.

Indiscriminate firing should be prohibited within the Yavuna area as well as the flanking highlands to safeguard against siltation-rapid runoff problems and the destruction of a valuable asset when the area is planted to trees.

An enlightened approach to fire danger through ground fuel accumulation following prohibition of burning must be adopted. It may be necessary to formulate a sound policy on controlled burning that will reduce fire hazard and yet achieve minimal damage to soil and vegetation. Inevitably this must take into account: (1) time of burning, (2) prevention of wildfire by patch and strip burning, and (3) protection of evergreen bush and plantations. Research will be required to provide these solutions.

## VEGETATION CHANGE IN THE NAUSORI HIGHLANDS

The Nausori Highlands comprise an 1,800 feet high plateau of complex volcanic rocks. Some magnificent forest remains. This is currently being milled by the Pacific Lumber Company which has a large, modern, integrated mill on the Highlands. Much of the former forest has been replaced by mission grass. This has resulted from firing. Runoff is rapid from the grassed areas. It is much less from forested areas. Stream flow is also better regulated from forested areas. Streams draining grassy areas rise quickly following rain. They carry off much fine topsoil that is swept downslope by sheetwash. Streams from the forested catchments carry a much smaller silt load. The heavy reddish clays found on much of this upland are relatively stable. They are deeply weathered and are not easily saturated, thus slips are less frequent than in the Yavuna area.

Experimental planting of British Honduras pine (*Pinus caribaea*), has shown that forestry is possible on the burned-over mission grass land. North American slash pine (*P. elliotii*), which has been planted at Nadarivatu, has proved better than *P. caribaea* at elevations above 1,500 feet. It would be a more suitable species for the Nausori Highlands.

Planned small farm schemes, fostered by the Fiji Department of Agriculture, have been developed in this area. One of the crops grown is seed potatoes as this area is virtually blight-free. There is evidence that a commonsense approach to rational land use is developing on the Nausori Highlands. If fires are kept out of the milled forests, natural regeneration will occur. The indigenous forest will be maintained. If a balance between planting of exotic conifers and development of agriculture could be developed on the mission grass areas the Nausori Highlands could become a very productive and diversified region.

## VEGETATION CHANGE IN THE BA REGION

The impact of non-Fijian cultures upon the environment is tremendous in the Ba region. Extensive areas of sugar cane cover the Ba Basin. Bush has disappeared from the periphery (Plate 7). The close link between firing, vegetation denudation and erosion in Western Viti Levu has been demonstrated nowhere more dramatically, and nowhere more disastrously, than in the Ba area. It is extremely difficult to visualize today that the denuded, bare, severely eroded hills and valleys of the Ba Closed Area, south of Ba township, ever supported *any* vegetation. That forest formerly clothed these areas appears unbelievable.

Frequent and repeated illegal burning was practised by the Indian cane growers of the adjacent lowlands. Firing, intended initially to provide grazing for working buffalo and oxen, eventually destroyed all the vegetation. Fires destroyed the bush. Subsequent, fire-induced *gasau*, scrub, grass, and fern communities each became increasingly depauperate with repeated burning. As vegetation cover deteriorated soils were stripped away from between the sparse vegetation. Fires and grazing eventually killed out even the fire adapted mission grass and the very hardy, xerophytic *qato*. With complete removal of vegetation, soils were stripped exposing bedrock over much of the area. Bare rock is widespread even now after several years of closure so that the now scarred landscape can begin to heal. Vegetation is virtually lacking. Hardy *qato* (*talasiga* vegetation), itself an indicator of bare eroded soils, is only beginning to colonize some of the less steep areas several years after the complete closure of this area to any land use or firing (Plate 8).

Siltation of the Ba river has proceeded rapidly following the erosion of soil from the flanking uplands. Runoff is extremely rapid from the eroded slopes and flooding has become an increasing hazard.

The impact of Indian culture upon the land in the Ba area initiated great changes. It has introduced severe problems. It will be many years before evidence of this reckless waste of a national resource is effaced. The severe erosion of the Ba Closed Area will long serve as a mute, but stark reminder of man's abuse of his environment.

## VEGETATION CHANGE IN THE DRASA AREA

Similar severe, though not quite so advanced, destruction of the land resource has also occurred at Drasa. Firing destroyed the bush cover and hastened the processes of laterization on the Tertiary lavas and agglomerates of this area. Frequent and repeated burning removed most of the vegetation and produced an eroded lateritic landscape sparsely covered in mission grass or *talasiga*. Twyford and Wright (1966) believe that removal of bush vegetation in areas with deep tropical weathering speeds the laterization process. Evidence for this is seen in the lateritic profiles now present on different levels at Drasa. The rapid formation of ironstone soils under sparse and frequently burned grass or fern vegetation precludes natural regeneration of these areas. *Nokonoko* are the only trees present.

Where fire has occurred within the last ten years sparse mission grass is present. In quadrats of one square metre, even on alluvial flats, only six to eight tufts of mission grass are present. Most of these tussocks are under three inches in diameter at ground level. On quadrats that had been burned recently two to three tussocks survived firing. This pattern is repeated throughout the area where burning has occurred. Occasionally a large tussock, up to 9 inches, across may be present. Many tussocks are only one inch in diameter. Seedlings (up to four per quadrat) are present on moist sites. Regeneration on burned slopes is minimal, usually in the form of two to three seedlings of mission grass. Many areas are completely bare. No signs of fern shoots are seen on burned slopes.

On areas of ironstone gravels *qato* is the only important plant cover. Often vegetation is completely lacking. Mission grass is generally absent.

In the Drasa area where fire has been excluded over recent years canopy cover of herbaceous vegetation is from 85-100 per cent. Basal or ground cover is much less beneath the grass tussocks. It is less than 20 per cent under the wiry *qato* fern fronds. Organic litter accumulation has begun. This cannot develop under frequent firing. Common species are *Dicranopteris linearis* (*qato* or *qato moce*), *Wickstroemia indica*, wire grass, mission grass, bracken, guava, *walutumailagi* (*Cassytha filiformis*), and *Desmodium* spp.

On sites where fires had been absent for several years the characteristic composition of several one square metre quadrats with 90 per cent canopy cover was: 300 stems of *qato*, 1 stem of bracken, 2 tussocks of mission grass, 1 shrub (usually *nuganuga*), 3 miscellaneous herbs (possibly *Desmodium* spp.).

With continued freedom from fires the plant composition alters. Grass and shrub vegetation slowly becomes more dense. Pockets of alluvial silt are trapped, *Desmodium* spp. become established and spread out from them.

Later, with ten or more years freedom from burning, trees or shrubs of *nokonoko, nuganuga*, and *wase* develop in the protection of the fern and grass. In the milder micro-environment of this pioneer community blue rat's tail thrives. A small silt-binding cucurbit with a net-like growth form is also common.

In 1955 small experimental plantings of 10 to 30 acres of British Honduras pine and various *Eucalyptus* were made at Drasa on eroded soils. Another 200 acres were planted later, and 500 acres were planted in 1966. This planting programme is to be expanded to 1,000 acres during 1968. *Pinus caribaea* grows rapidly: it is slightly faster than *P. radiata*. It is also a relatively fire resistant species. If fires are kept from the Drasa area pine plantations can be established even on severely eroded and lateritic soils. This type of land use will at least maintain the permanent productive capacity of this area. Evidence is present that afforestation will not only arrest the deterioration of the Drasa area but will gradually increase the permanent productive capacity of the area. A return to some form of bush cover is highly desirable. It will regulate runoff, retard erosion and provide an economic product. Firing, with its resultant mission grass and *talasiga* vegetation and aggravated erosion cannot provide any of these.

The following patterns of vegetation were observed in the Drasa area under planted trees. Fire had been excluded for various periods. Ground cover on areas of ironstone gravels with established four year or older *Eucalyptus* was chiefly *qato*. When unburned, it provides a three-tiered canopy. The upper living fronds provide the top layer; the previous season's fronds form an erect secondary layer; and previous seasons' growth, which is dead and semi-prostrate, the lowest third layer. Together these greatly reduce the physical impact of rain. Interception is quite high. These layers of the *qato* fern result in a reduction in temperature range compared with open and bare areas. There is an abrupt boundary between dead organic litter and the ironstone gravelly soils. There is no humus mat: the vegetation does not act as a sponge controlling infiltration but it does reduce surface impact and runoff. After two years seedlings of *nuganuga* and *wase* may appear in the milder micro-climate of the fern cover.

Under six-year-old pine stands on gravelly lateritic soils where mission grass replaced *qato* the pine needles formed a shallow two-layered mat of (a) undecomposed and (b) semi-decomposed needles. An abrupt boundary between organic litter and mineral soil was again present. However the ¾ inch pine needle mat greatly reduced surface runoff and erosion and also served as an insulating layer modifying temperature ranges at the soil surface.

Runoff observed during high intensity rains from (1) bare areas (i.e. recently burned), (2) fern-covered but eroded and burned earlier, (3) unburned eroded areas with one year pine seedlings, (4) unburned with two-year growth, and (5) unburned areas with more than two years' growth gave the following results: With the increasing age of the plant cover there was an obvious trend for decrease in (a) rapidity of runoff, (b) turbidity of water, (c) size of particles transported, (d) volume of particles transported and (e) velocity of stream rivulets. Flow of water was negligible from the older vegetation areas and the water was clear.

## VEGETATION CHANGE IN THE TAVUA BASIN

Mission grass covers the slopes from the cane growing areas to the mountain crests and beyond (Plate 4).

Indian cane farming has penetrated higher up the northern slopes of the Tavua basin than elsewhere in western Viti Levu. Firing is very common. Fires occur frequently and are very widespread. Mission grass fires often rage for miles and sometimes burn for days. A blue pall of smoke and a coppery sun are commonplace by day. By night they are replaced by twinkling lines of fire creeping up the mountain slopes. Some guava shrub is present but it is not as widespread as elsewhere in western Viti Levu.

Savannah-type light forest is present as a narrow transitional belt on slopes below undisturbed forest. It appears where burning has not been frequent after original bush has been burned. Trees are often low but range from 20-45 feet in height. They vary from widely spaced to more dense. Scattered shrubs occur as a dense under-storey with grass, reeds and herbs.

Common trees are *drou, vasa* (*Cerbera odollam*), *quaqua* (*Conthium rectinervium*), *tadano* (*Homalanthus nutans*), *davo—a* species of *Macaranga, yaro* (*Premna taitensis*), species of *Ficus*, and *koka* or *koka damu* (*Bischoffia javanica*). Guava may be present as a tree or a shrub.

Shrubs present include guava (*Glochidion anfractuosum*), *dranigata* (*Leucosyke corymbulosa*), *nuganuga* (*Desmodium heterocarpum*), *Acalypha repanda* var. *denudata* and *Wickstroemia indica*.

*Gasau*, mission grass, *Sida* spp., *Polygala* spp., peanut 'grass', Chinese burr, *Desmodium* spp., tar weed, *qiriqiri* ('rattle-pod') and various other introduced weeds form a dense herbaceous cover. If these areas are burned the community becomes reduced to one dominated by guava and mission grass. Mission grass dominates most of the hill slopes flanking the Tavua basin.

Legislation has been passed prohibiting illegal firing of grasslands. There is provision for fines and penalties if offenders are caught. The major problem is the difficulty of enforcing the legislation. Another problem is the extreme difficulty of attempting to put out fires on the steep terrain. The chief reaction to a fire in this area is to shrug and look the other way. Unfortunately this attitude is generally present elsewhere on the island.

The general problems of fire, vegetation denudation, runoff, and erosion, seen elsewhere in dry zone Viti Levu, hold true in this area. Some very severe floods have caused extensive inundation of parts of the Tavua basin cane lands.

The Fiji Forestry Department has an ambitious programme of afforestation for this area. Planting of *P. caribaea* on the lower, steep, mission grass slopes at Koro No. 1 and of *P. elliotii* at higher elevations near Nadarivatu is an attempt to return the mountain flanks of this productive lowland basin to some form of bush cover. If successful it will provide an important alternative source of income for the Colony.

Problems of fire control and management, which were outlined earlier when discussing possible afforestation in the Yavuna area, apply here. One can only hope that this commonsense attitude to land use—the most recent phase in man's attempt to modify his environment—will be successful. It is the closest approach to the pre-European Fijian's conservation of the basic resources of his environment.

## VEGETATION CHANGE IN THE LEWA AREA

Inroads by fire have been very deep in the high ranges inland from Nadarivatu. Dense mixed forest, essentially similar in character to the forests of the wet zone, have been replaced, through firing, by savannah scrubs, guava, and mission grass vegetation. In some cases even the mission grass has gone and bare eroded slopes remain where once dense forest stood only decades before.

Erosion has not been as severe in the high Lewa area as in the Ba Closed Area. However, widespread sheetwash is prevalent under mission grass in this high rainfall area. Mass movement processes of slumping and slipping are common in the saturated Lewa clays on non-bush areas near Tumbeilawa and Naidadara valleys. Evidence of similar mass movement processes are not obvious in the adjacent bush areas. Stripping of surface soil horizons was noticeable on mission grass areas east of Korotevu and beyond Louokomburu.

Even within four years young plantations of *P. elliotii*, planted near Lewa, showed a considerable improvement in the catchment qualities of the areas where they were planted. This paralleled observations made at Drasa. The tree needles and the accumulated grass litter reduce raindrop impact on the soil. Surface runoff is much less and slower. Infiltration capacity is greater than on adjacent mission grass slopes. With time the differential between these will become more marked.

Measurements during a short period of field investigations showed that between 60 and 70 per cent of rainfall falling on the mission grass was lost as surface runoff. From 5-10 per cent was lost from the pine plantations. Only a trace of measurable runoff was recorded for the same heavy intensity rains in the bush. Rate of infiltration, in inches per hour, was from 0·5-1·5 for eroded mission grass areas, 2·0 for dense mission grass-guava scrub (unburned for several years), 2·2-2·5 for the young pine plantation and over 5 for undisturbed forest.

If fire can be kept out of this area its return to a form of bush cover will be rapid. There will be a great improvement in the catchment qualities of such areas. Both surface runoff and erosion will be reduced with a return to tree vegetation.

## CONCLUSION

There is no reason to believe that the studies of runoff and soil loss, water holding capacity, infiltration, and interception, under different types of vegetation made by Campbell (1945a, 1945b) in New Zealand are not equally true in Fiji. Preliminary investigations and measurements have indicated a close parallel for similar vegetation and soil conditions. Geiger (1965) has also demonstrated these fundamental truths.

Man's activities in western Viti Levu have resulted in many changes in the vegetation. Many problems have resulted. Over wide areas the basic principles of conservation have been disregarded. Often they have been flagrantly flouted. A more commonsense attitude to the use of the resources of the environment is obvious in planning today. However, the filtering process from administration, through the various channels, to the Indian cane grower and to the Fijian villager may be very slow. Much of the future pattern of vegetation change and of its problems depends upon both the speed and spread of this filtration process.

Three short sentences introduced this essay. It is fitting that they end it also: 'Neither man nor the environment is static' . . . This has been dramatically demonstrated in western Viti Levu. 'Human effort ebbs and flows' . . . and with it, vegetation frontiers have advanced or retreated. The spread of fire-induced savannah grasslands in western Viti Levu has been most vigorous and widespread over the last forty years. It has been closely related to intensive commercial cultivation of sugar cane essentially by Indian farmers. The next twenty years may show an equally important change if fire is controlled. Present grasslands may be replaced naturally by scrub forests in the absence of fire. By man's effort, afforestation—which shows early promise—may prosper. Extensive, dark, conifer plantations may replace present brown grasslands. These, and other changes not listed, will create a new set of environmental conditions. Thus will 'environmental factors appear and disappear'.

## ACKNOWLEDGEMENTS

Partial financial assistance for field work was provided by the University of Auckland South Pacific Programme Golden Kiwi Research Grant. The author wishes to thank various officers of the Fijian Departments of Forestry and Agriculture for their technical assistance and critical discussion, and Mr Reid, Pacific Lumber Company, Nausori Highlands, for his hospitality.

## BIBLIOGRAPHY

BARTHOLOMEW, R. W. (1960). *Geology of the Nadi Area, Western Viti Levu.* Geol. Surv. Dept. Bull. 7, Govt. Press, Suva.

BATCHELDOR, ROBERT B., AND HIRT, H. F. (1966). *Fire in Tropical Forests and Grasslands.* U.S. Army Tech. Report 67-41-ES, Mass.

BURNS, A. *et al.* (1960). *Report of the Commission of Enquiry into the Natural Resources and Population Trends of the Colony of Fiji*, 1959. C.P. 1/1960.

CAMPBELL, D. A. (1945a). 'Investigations into runoff and soil loss', *N.Z. J. Sci. Tech.*, **A26 (6)**: 301-32.

CAMPBELL, D. A. (1945b). 'Investigations into soil erosion and flooding', *N.Z. J. Sci. Tech.*, **A27 (2)**: 147-72.

COCHRANE, G. ROSS (1967). 'Land Use Problems in the Dry Zone of Viti Levu', *Proc. 5th N.Z. Geog. Soc. Conf.* 111-118.

C.P. 1950 *et seq.* Departments of Agriculture and Forestry Reports in Journals of Legislative Council of Fiji (*Council Papers*).

C.P. 11/1966. Fiji Development Plan.

C.P. 16/1966. Fiji Development Plan.

DERRICK, R. A. (1957). *The Fiji Islands.* Govt. Press, Suva.

FRAZER, R. M. (1961). *Fiji.* CAB, **27**: 6. Sydney.

FRAZER, R. M. (1962). Land use and population in Ra Province, Fiji. Unpublished Ph.D. Thesis, A.N.U., Canberra.

FRAZER, R. M. (1964). 'Changing Fijian Agriculture', *Austr. Geogr.* **9**: 148-55.

GEIGER, R. (1965). *The Climate Near the Ground.* Harvard Uni. Press, Mass.

MARSH, B. (1965). Soil Erosion, Nadi Hinterland. Dept. Agric., Fiji. S8/1, 24-7-65 (Unpublished Report).

PARHAM, J. W. (1955). *The Grasses of Fiji.* Dept. Agric. Fiji, Bull. No. 30, Suva.

PARHAM, J. W. (1958). *The Weeds of Fiji.* Dept. Agric. Fiji, Bull. No. 35, Suva.

PARHAM, J. W. (1964). *Plants of the Fiji Islands.* Dept. Agric. Fiji, Suva.

PRICE, A. GRENFELL (1939). *White Settlers in the Tropics.* American Geographical Society Special Publication, No. 23, New York.

SEEMANN, B. (1862). *Viti: An Account of a Government Mission to the Vitian or Fijian Islands in the years 1860-61.* Macmillan, Cambridge.

SEEMANN, B. (1865-73). *Flora Vitiensis: A Description of the Plants of the Viti or Fiji Islands with an Account of their History, Uses and Properties.* Reeve, London.

SIMONETT, DAVID S. (1967). 'Landslide distribution and earthquakes in the Bewani and Toricelli Mountains, New Guinea', in J. N. Jennings and J. W. Mabbutt (eds) *Landform Studies from Australia and New Guinea.* A.N.U. Press, Canberra.

SPATE, O. H. K. (1959). *The Fijian People: Economic Problems and Prospects.* Fijian Legislative Council Paper, C.P. 13/1959.

TWYFORD, I. AND WRIGHT, A. S. C. (1966). *Soils of Fiji.* Govt. Press, Suva.

WARD, R. GERARD (1959). 'The Population of Fiji', *Geogr. Rev.*, **49**: 322-41.

WARD, R. GERARD (1960). 'Village Agriculture in Viti Levu, Fiji', *New Zealand Geogr.*, **16**: 33-56.

WARD, R. GERARD (1965). *Land Use and Population in Fiji.* Dept. Technical Co-operation, Overseas Research Publication No. 9, H.M.S.O., London.

BRIAN MAEGRAITH

# *Jet Age Medical Geography*

*'. . . disease has played and is playing a vital, if little realized, part in world events.'*

A. GRENFELL PRICE *The Western Invasions of the Pacific and its Continents*

## HUMAN MEDICAL ECOLOGY

SPEAKING OF human medical ecology, which in its broadest sense embraces the study of the inter-reaction in both directions of man and his environment, Jacques May laid it down that what we want to know is—who has what, where, and why?

Dudley Stamp has pointed out that setting out the facts as we at present know them is partly a matter of devising methods of providing the requisite distribution maps—cartograms, as he calls them—ranging from world maps to plans of countries, towns, villages and even houses. Such charts should be regarded as research tools. They represent static points in a continuously progressive process. In some circumstances a map may reveal an otherwise hidden factor in the spread of disease, as in Snow's map of cholera deaths in London in 1848, which pin-pointed the role of contaminated water in the spread of *Vibrio cholerae*. It may indicate the spread of disease as a result of pacification or settlement of specific areas of a continent. It may define the effects of war on man and animals alike as in the spread of plague in South Vietnam today.

In fact, the cartogram is usually out of date before it is drawn. Man has long shown himself capable of adjusting his environment to suit himself. He can move singly or in masses from one area to another; he can adjust himself and his life to new environments. This has been his history and has slowly been changing the concept of the geography of his life and of his health.

The technical revolution of this century has incredibly increased the pace of these changes. It is now possible, for instance, with air conditioning, to change a local environment to suit the living conditions and physical health of expatriates in hot climates, as in Kuwait. By modern irrigation methods or by aerial replacement of trace-elements, deserts can be made fertile, as in Egypt and the Sudan, in Israel and Australia.

Oceans have been joined by canals to facilitate the movement of larger and faster ships. In the last two decades distant places have been linked by the aeroplane which, on a commercial basis, can take a man halfway round the world inside two days and will soon carry him from Europe to the Americas in the same number of hours.

Similar processes are in action in the so-called emerging countries. Plans are being mounted for vast socio-economic developments. Dams are being built for supplying electricity and for irrigating land for food or cash crops. Rivers are being diverted for land reclamation. Roads and services are being taken to remote areas of the jungle and desert. Man is rapidly changing his environment and, by the consequent movement of labour forces, by resettlements and introduction of new farming and industrial populations and their products, is re-shaping not only the old-fashioned geographical anatomy of his local world but the social and physical environment in which he has to live.

He is also changing the biological environment. Animals, reservoirs of disease, are being driven from one area to settle in another, taking with them organisms which may infect new human and sometimes non-immune populations. Insect vectors are losing ground in one area and gaining in another. Man's habits of life are changing with the changing conditions and exposing him to new health hazards.

All these things mean new people in new environments, Caucasians living in the sun, tropical peoples in the cold north. Individual movement is now completely fluid. The migration of communities is already creating world problems. The patterns of disease are themselves being changed and the problems of man against his environment are altering in respect of place, and time and intensity.

Disease has no political boundaries but nations have, and this in itself greatly influences man's own attempts to tackle his health problems both locally and at global level.

## POPULATION MOVEMENTS

Man seems to be the only animal capable of creating its own ecology. As he develops scientifically and technologically, and changes and disturbs his environment, he also affects his own reactions to it.

Sometimes he leaves an inimical environment and seeks another. There are many historical examples of this. Disease has caused many migrations, just as the population movements themselves, as in a world war, have temporarily or permanently disturbed ancient patterns of disease distribution.

Thus, areas which were otherwise rich and inviting have been emptied of their human population by disease, for example by malaria in India, by river blindness in northern Ghana, and by trypanosomiasis in East Africa.

Diseases which need only human contact for their propagation have had great influence on the history of human geography. They may be introduced into a relatively static community by a few immigrants, as was smallpox into the new world in the sixteenth century, reportedly by a single Negro. In this case, the epidemics which followed accounted for millions of deaths.

The spread of some infections which have had devastating and prolonged influence on populations have been dependent on whether the specific vector or the intermediate hosts were available. Thus infections indigenous to the tropics have become established in new areas of the world where the insect vector or the intermediate hosts existed although the disease itself did not exist or had been eradicated. Probably the best known example of this is the introduction and establishment of schistosomiasis and malaria in the Caribbean and the Americas following the forced mass migration of infected slaves from Africa. It is now believed that filariasis, guinea worm infection, onchocerciasis and possibly yellow fever were also introduced into this region in this way.

Sometimes it is not the human migration that counts, but that of the vector. A most remarkable example of this was the introduction into north-east Brazil in 1929-30 of the malaria vector *Anopheles gambiae*, presumably carried across the Atlantic by a fast French destroyer from Dakar in Senegal. In Brazil the mosquito, which is an extremely good vector of malignant malaria and was previously unknown in the region, found no natural enemies and multiplied and spread prodigiously. Severe outbreaks of malaria began in 1931 and over the next seven years accounted for 100,000 cases and 13,000 deaths. This deadly visitor to Brazil was eventually eradicated by Soper (1943).

Other movements of populations have had disastrous effects on indigenous peoples, not only physically but economically, sometimes interfering with the growth and development of whole nations. Temporary migrations forced by military occupations, the introduction of infected troops into non-immune populations and the reverse have influenced the outcome of wars and revolutions. Construction labour forces concerned with building roads and dams have changed the patterns of infections inside developing countries. Sometimes, as in the original attempts to build the Panama canal, local disease has put a complete stop to progress. Recurrent migrations involving very large numbers of people moving into special areas, such as the pilgrimages to Mecca and to the holy places in India, have provided health hazards both to themselves and the local people.

One form of population movement which has had and still has a profound effect on the distribution of disease in Asia and in Africa in particular is the cyclical migration of nomads such as the Fulani in search of water and pastures for their animals. Prothero (1965) has pointed out that in Africa the movements of true nomads (as distinct from semi-nomads who return each year to some focal point and cultivate a minimum of crops) are difficult to define. It may be possible to distinguish some sort of annual cycle but the routes can rarely be detailed with precision, since they depend on when and where rain falls and may change from season to season. The nomads 'rank among people least touched by political, social and economic progress in mid-twentieth century Africa'. In many newly independent countries attempting the long haul to economic and social progress they amount to an anachronism. Apart from the hazards of transmission of infections carried by them to the inhabitants of the territories through which they pass and their own exposure to endemic infections, they are at the very least a major nuisance in nation-wide campaigns against disease. Indeed, failure of local eradication operations may sometimes be fairly attributed to seasonal movements of nomads and of migrant labour.

Such migratory movements are traditional in Africa and will continue. Eradication programmes will have to be adapted to them, and will be consequently slowed. This will affect the whole socio-economic development of the rural areas of the continent and, with the other human problems that must be faced in any attempt to control disease, will influence the medical geography of Africa for a long while to come.

Prothero has noted that the appearance on the African scene of the European powers led to political boundaries being drawn in many places across ethnic groups. These were largely ignored by the peoples concerned until very recently. The majority of these former colonial territories are now independent States, but the same borders have been kept and are often more rigidly controlled than before. One effect has been the discarding of traditional migration routes in favour of new ones, sometimes outside the continent. This has led to a corresponding redistribution of disease patterns and often failure of collaboration between contiguous countries in the attempted control of major endemic diseases.

Artificial political boundaries drawn across ethnic groups are difficult enough to maintain even with military operations. Where they are also drawn across endemic areas of vector-borne disease the chances of co-operation in a common effort in eradication are often slender indeed. This has been seen again and again in world malaria eradication programmes: in Asia on the Thai-Burmese and Chinese borders where control on one side has not been matched on the other; in the Venezuela–Colombia borders where difficulties of terrain are immense and where raiders have from time to time rendered any attempt at proper execution of an eradication manoeuvre almost impossible and sometimes lethal; in West Africa where Ghana abuts on Togo and the Ivory Coast.

Large-scale movements of populations are not always needed to disturb the delicate balance between health and disease in a community. A few infected individuals entering an area in which a specific infection does not exist, although its natural vector does, may cause devastating epidemics. Earlier this century, the introduction of infections commonly occurred as a result of extension of trade and the peaceful settlement of formerly unsettled peoples. Today, as a result of political unrest and war, trypanosomiasis is spreading in areas in which it had been controlled in the Congo and in Nigeria.

There is always some social factor somewhere upsetting the relationship between man and his infections and it is useful in this context to realize that when we talk about the distribution of disease, we are normally thinking in terms of the human host, parasite reaction and all that this represents, although we often loosely talk only in terms of the infective agent or its vector; for instance, the distribution of the Trypanosome and the tsetse fly. As Henschen (1966) has succinctly put it: 'Diseases are not living things; they are not concrete phenomena which can exist apart from mankind. Diseases in themselves are abstractions; in reality there are only sick men'.

Perhaps it might have been better to say 'infected' rather than 'sick', but curiously enough, in many of the conditions occurring in man today it is not always easy to distinguish between an infection and a disease. An example of this is amoebiasis. Should a man be regarded as 'diseased' if he is carrying a commensal *Entamoeba histolytica* infection and passing cysts? Should cyst passers be included in an epidemiological survey of the disease as such? There are many interesting points of this sort to be solved by the geographer in collaboration with the doctor. Failure to understand the medical implications of available information has led to justifiable suspicion of some of the work presented by the lay geographer, since the sources of information are often extremely unreliable or need technical interpretation.

Migrations under social or political duress of populations from one part of a continent to another have caused direct exacerbations of existing infections or have infected local vectors which were previously uninfected. The very severe outbreaks of trypanosomiasis in northern Uganda in the early days of this century arose in this way following enforced resettlement as a result of an epidemic of the same disease in their original location. These were in themselves a tremendous set-back to local developments. Similar spread of infection, in this case of schistosomiasis, followed the more recent resettlement of Portuguese into the Limpopo River Valley Irrigation Scheme in Mozambique.

Unless precautions are taken, such a calamity may occur again wherever labour forces move within a country to work on socio-economic ventures such as dams and roads and wherever settlers are placed in new areas in which the local infections have not been controlled.

In terms of human misery and frustrated attempts towards economic and social progress, infection is often an avoidable risk. In view of the tremendous upsurge of socio-economic development in emerging countries today the danger of spreading infection should therefore be faced realistically in the stage of planning and not tackled when disaster is imminent or actual. This seems to be a lesson that many governments are finding hard to learn, although there have been exceptions, as in the building of the Kariba Dam. In north-east Thailand, the Pong-Neep dam (now called the Ubolratana Dam) was finished last year and is now functioning and providing electricity over a wide area of the country. The dam itself was constructed by one group of expatriate nationals, another made the roads and built the bridges, others did the requisite geological and hydrological surveys both for the hydro-electric dam and a lower dam which will eventually provide for the irrigation of some 250,000 acres for rice growing. No one except the local Ministry of Public Health was, in the early stages, interested in the health problems likely to be posed. This was pointed out and now the whole situation has been improved by the creation of a national council for dealing with the health situation. Today the overall plan is continuing in parallel with surveys of the endemic diseases present in the relevant areas and the necessary public health measures are being put into practice before the influx of new populations into the areas to be irrigated.

This approach is an elementary blue-print for all socio-economic planning in emerging countries. Social and economic set-backs and disappointments which follow the introduction or exacerbation of disease in areas of development have been demonstrated too often. And where they have occurred, the progress of the people, sometimes of whole nations, has been retarded. In respect of schistosomiasis alone this has been seen in Egypt, in the Gezira in the Sudan, in the Yangtse Valley in China.

Perhaps the most remarkable achievement in the opposite direction has been the control of disease and the consequent acquisition of high standards of health and living in modern Israel, itself a remarkable example of mass migration. The success of Israel has largely depended on its internal health measures and the control of disease amongst immigrants from all over the world, many of whom in the early days were infected with tuberculosis or brought in schistosomiasis from local African and Middle East States. Strict control of infection, and improvement in health standards have been major factors in the success and survival of Israel in its otherwise inhospitable international and physical environment. In this respect it is interesting to note that this has been achieved without co-operation in disease control across national boundaries. Rather the reverse, since, by firmly shutting her frontiers, Israel has avoided the dangers of the population movements, including the cyclical pilgrimage migrations to Mecca, which disturb all the contiguous countries.

## INTERNATIONAL AND NATIONAL DISEASE CONTROL

The medical geographer attempting to draw his cartogram of modern disease distribution will have to take into full account national and international attempts to confine disease.

International sanitary regulations in the form laid down by the World Health Assembly (1951-60) have had a very considerable effect on the reduction of the world incidence of the officially quarantinable diseases (plague, cholera, yellow fever, smallpox, typhus and relapsing fever) and have thus influenced the spread of disease by international travel. As will be seen later, however, these regulations are based on the time schedules of sea travel and not those of air travel. Because of the speed of the latter, there has been some failure of international control, since disease may be introduced by individuals during the silent incubation period of an infection or by unidentifiable carriers of infection. It is difficult to see how international regulations can legislate for this. In any case, there is clearly need for review of the present situation, as Dr Candau, the Director-General of the World Health Organization recently emphasized (1966): 'There is little doubt that some of the traditional national and international control measures and regulations may soon become outmoded.'

An example of the possibilities of this failure (we hope, temporary) of international and national control of the spread of infection is the dissemination of cholera.

Until recently the organisms responsible for classical cholera were the three principal strains of *Vibrio cholerae*, one of which usually predominated in any given outbreak or epidemic. A fourth vibrio (El Tor) was known to exist in certain areas of South-east Asia but was not regarded as potentially dangerous.

Since the classical vibrios do not survive for long in the infected patients, 'carriers' of them in the sense of typhoid carriers do not exist and the threat of global spread, even by aeroplane, is thus minimal. Until the last decade, therefore, it was generally considered that the disease was under control at an international level, except in the areas of India and the Far East where it was endemic.

The picture has radically changed for the worse. The El Tor organism has now caused widespread outbreaks of severe and often fatal cholera in South-east Asia and India and in many areas has replaced the classical forms of the organism. Moreover, the El Tor vibrio may persist for long periods in the excreta of infected persons, whether they have had the clinical disease or not. True carriers of the infection thus exist, and constitute a real menace to global health, since they could theoretically carry the infection anywhere and cause outbreaks wherever the local public health standards, especially in regard to water, are low. This is already happening, as can be seen in the appearance of isolated El Tor cholera cases or outbreaks in places as far away from the endemic areas as the countries of the Arabian Gulf. Next time it could be eastern Europe or remote Wales.

We must take note of the courageous lead by the World Health Organization in introducing the concept of eradication of a disease from the face of the globe. The idea of eradication of malaria which at one time threatened almost two-thirds of the world population is remarkable enough and the achievements to date in some areas are startling. In the few years since the project was begun vast areas have been cleared of the disease. Nevertheless, the plan has been slowed in some areas and progress in many difficult regions is discouraging. This is hardly surprising since it has demanded the elimination of the vector mosquitoes from all kinds of terrain and the eventual chemotherapeutic destruction of the parasites.

Unexpected biological hazards have appeared which have restricted progress. Mosquitoes have become resistant to residual insecticides. Parasites resistant to all the usual synthetic antimalarial drugs and responding only to quinine have appeared and spread in South-east Asia and South America.

There have been social and political difficulties too, the latter in particular. Major causes of failure have been population mobility, incomplete geographical reconnaissance before commencing operations, difficulties in communication and the extremely variable socio-economic conditions of the communities in which the disease exists. It has become obvious that eradication of malaria from a country cannot be achieved unless the local government can afford the cost of the necessary surveillance, a matter which involves the provision of both budget and trained personnel. There are few developing countries in the malarial belts of the world, especially in Africa, which can afford this even on a very long-term basis. Moreover, eradication is impossible if contiguous countries cannot or will not co-operate. Mosquitoes are not intimidated by customs barriers.

Thus, progress is unhappily slowed, but nevertheless big changes have already been made in malaria distribution and the world map of the disease has been considerably redrawn. In circumscribed geographical areas, notably Cyprus and Sardinia, eradication of the disease, if not the vectors, has been complete. The effect on the local communities has yet to be fully demonstrated but there is little doubt that lifting the burden of malaria should raise standards of health and provide opportunities for improvement in living standards.

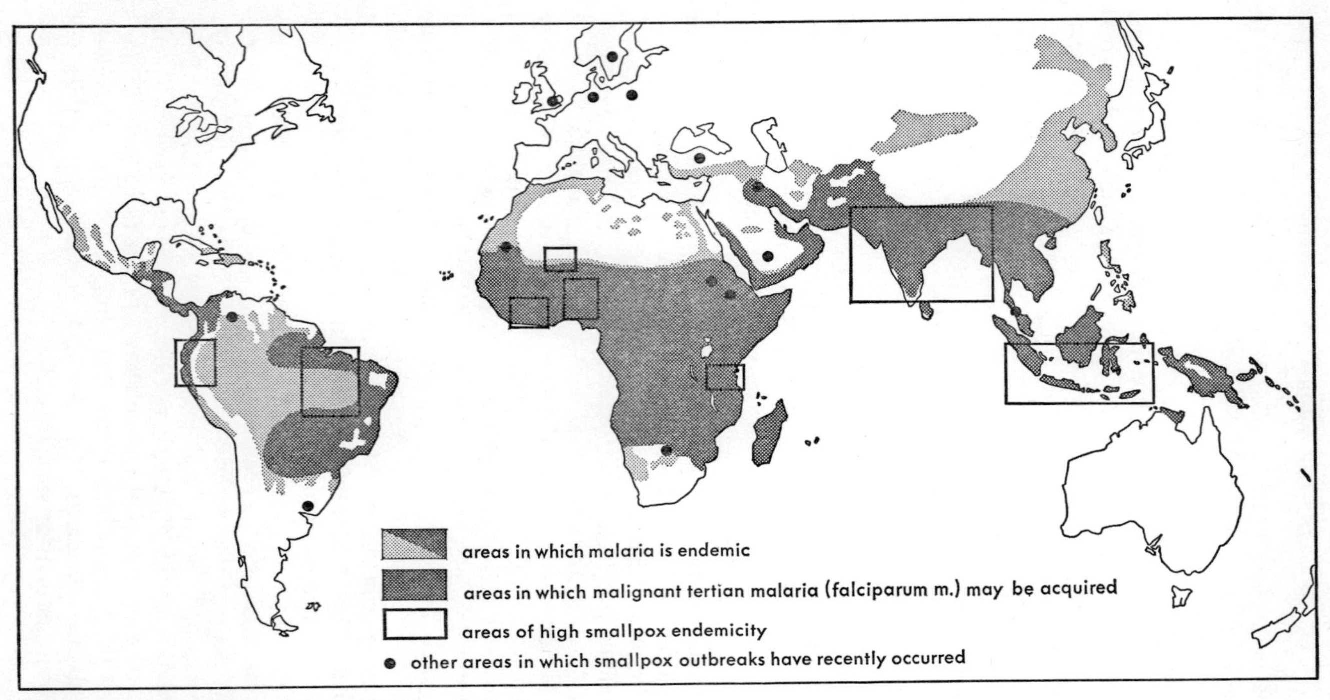


FIG. 24 World distribution of malaria and smallpox

So man is consciously changing not only his local environment but the whole of his world. Malaria eradication is a goal not yet nearly reached, but the concept has changed the philosophy of disease control. One day malaria may be confined and even eradicated. Why not apply this aim to other conditions, such as tuberculosis? It has already worked extremely well in the case of yaws in Africa and poliomyelitis in Australia.

## MEDICAL EFFECTS OF MIGRATIONS

Charting infections in themselves is clearly not enough in considering human medical geography. The distribution of *Leptospira* in rodents in areas of north-east Thailand scheduled for irrigation gives an indication of the potential menace of the disease in the human population. Nevertheless, only figures of the incidence in man can declare the human infection and disease situation. Such information is often extremely difficult to collect. Maps of the distribution of infectious diseases which are essentially zoonoses or in which man and animals are both definitive hosts are commonly compounded of information from both sources and may therefore be suspect.

This, however, is only one hazard in deciphering geographical medical data, since genetic factors in the hosts themselves may play an important role in the development of human infection and disease.

The supreme example in the tropics is the abnormal haemoglobin S, which is transmitted by mendelian inheritance and which may thus appear in siblings in the heterozygous or homozygous state. It has been shown in heavy infection with *Plasmodium falciparum* (malignant tertian malaria) that the heterozygote (i.e. the individual with the S trait) resists the serious consequences of the disease and survives where the individual with normal haemoglobin dies. This results in human selection, which is going on under our eyes at this moment in Africa and elsewhere where the gene has been taken, in particular by the slave trade to the Caribbean and South and Central America.

Simple maps of the distribution of malaria in these areas are therefore not enough in terms of human morbidity and mortality, since the trait in itself does not affect the incidence of the infection.

There are many other examples. There is some evidence that red blood cell deficiency in the enzyme glucose-6-phosphate dehydrogenase may offer similar advantages against *P. falciparum* malaria, whereas many of the abnormal haemoglobins other than

S do not. Again, some forms of thalassaemia interact with either S or E haemoglobin, but not with other abnormal haemoglobins, to cause diseases which are severe and often fatal in childhood. Hence the consequences of thalassaemia itself vary in a strictly geographical manner, since S haemoglobin occurs largely in Africa and in the old slave-trade areas and E haemoglobin exists only in the East.

It will be noted that the distribution of haemoglobin S in the new world more or less follows the pattern of the forced movements of the slaves taken from the West African coast to the Americas. This is a nice example of a medical geographical pattern being dependent essentially on migration, a factor which, as we will see, still influences the distribution of disease.

## PROBLEMS OF NUTRITION

The medical history of man, which in many ways is also his geographical history, has not been shaped only by infection. All aspects of his health have been involved. Few things have been more important than food and water. Where these have been plentiful people have settled and prospered unless exposed to other factors such as infections. Where there have been shortages, emigration to better conditions has often followed.

Under-nourishment and malnourishment are still the lot of the bulk of mankind and must be considered in discussing the effects of disease as such on his distribution and ecology. Indeed, the word 'disease' includes the results of both infection and of malnutrition.

In such circumstances the brunt of the attack commonly falls on the child and this in itself has a profound effect on the pattern of world disease and population. Whatever the race or creed, where people are crowded and food is short and particularly where the diet is badly balanced, with a staple rich in carbohydrate but poor in protein, kwashiorkor and marasmus appear and children die. This is true not only of rural populations. It also occurs in the towns and even in big cities such as Bangkok, where food supplies are adequate but social conditions are not, where mothers, for economic reasons, return to work immediately after their babies are born and feed them on thin mixtures of tinned milk and water rather than on the breast. Here the fault is largely a social one compounded of economic pressure and ignorance.

Nevertheless, most nutritional troubles are still seen in the poorer parts of the world, especially in the so-called tropical countries. Nutritional requirements in the tropics are basically the same as in temperate regions, but the diet of native populations is largely governed by custom and supply and is usually based on some staple carbohydrate foodstuff, such as rice, to which other substances are added almost fortuitously. These diets are deficient in high grade protein and often in fat and essential substances including vitamins and minerals. It is not surprising therefore that malnutrition and other nutritional disturbances are widespread. Apart from their own intrinsic effects on the population, these conditions form a background against which the common infectious diseases develop, and the latter, in turn, may complicate the issue by interfering with intestinal absorption and even with the metabolism of foodstuffs.

It is often said that malnutrition tends to exacerbate other disease processes, but such statements are not always justified. It so happens that the worst ravages of malaria are usually found in malnourished and poverty-stricken communities and the conclusion is frequently drawn that the malnourishment itself contributes to the results of the infection. This view contrasts sharply with recent observations in western Nigeria, where it has been shown that fatal cerebral malaria is commoner in well-fed than in malnourished children suffering from kwashiorkor. Moreover, there is experimental evidence which indicates that in a starving animal or in animals living exclusively on milk (and thus short of para-aminobenzoic acid), the development of malaria parasites is inhibited, not accelerated. The truth is that we do not yet know much about the interaction of malnutrition and disease, and until we do, it will be difficult to equate the two factors in man's medical geography.

## MAN, ANIMALS AND HABITS

A factor of great importance in relation to human disease arising from both infection and malnutrition is disease in animals, since in the tropics where man and his animals often live so closely together, many infections involve both, and man is dependent essentially on animals for his high grade protein foods.

Some virus diseases are primarily infections of animals accidentally transmitted to man. Thus yellow fever is enzootic in many animals in the forests of Africa and South America and may be confined for long periods to the tree tops, reaching man through infected monkeys raiding his orchards or deserting the forests when they are cut down. In this way the disease may appear first in the small jungle villages and spread to bigger communities as the local *Aedes* mosquitoes become infected.

Worms, such as *Schistosoma japonicum*, may infect domestic animals as well as man and so help distribute the infection amongst rural workers, as in the rice fields of the Yangtse Valley. Many of the worm infections are spread by dirty habits of living. The life cycle of the infective agent in schistosomiasis, for instance, begins when an infected individual passes faeces or urine containing the eggs into water in which the intermediate snail host is living. The extension of the infection and its persistence in a given area, perhaps coloured pink in the geographer's map, depends therefore on the habits of the population. The colour indicates not only the distribution of the infection; it reflects also the habits and the standards of health and living of the local population.

In some instances the life cycle of a parasite and the spread of infection are dependent on the local animal population or on local human habits of feeding. Liver fluke infection in north-east Thailand results from eating raw fish infected with metacercariae of *Opisthorchis*, lung-fluke infection in Asia and Africa from eating raw crab meat and intestinal fluke infection from eating certain vegetables. A map showing the distribution of opisthorchiasis, paragonimiasis or fasciolopsiasis in Thailand will thus also demonstrate the distribution of the habit of eating raw fish, raw crab or water caltrop.

Removal of the colour from the particular area on the map, representing the disappearance of an infection, could therefore also indicate a change for the better in human habits. This is rarely the case. It more often means that a successful (unfortunately often also temporary) way has been found to break the life cycle of the parasite rather than disturb the life habits of its definitive host.

There is little doubt that control of such infections would improve local health and living conditions, and it is these factors just as much as the presence of the disease itself which must concern the medical geographer. In the long run, even a chart of a vicious epidemic is no more than a contemporary map of the state of health of the community involved.

## THE JET AGE

The modern medical geographer is in the happy position of having it both ways.

In many parts of the world the older processes are continuing unchanged or are being slowly and sometimes violently challenged by developments. New-found independence and immense plans for socio-economic development, particularly in Africa, are rapidly changing the social and economic structure of whole communities.

So far, we have discussed general factors which have helped draw the maps of medical geography in the past and are involved today. Many of these are basically traditional and will continue for a long while to come to influence life and health in the vast rural and agricultural communities of the primitive world. But the impact of modern developments is already being felt even here, and the scene is changing in a bewildering mixture of speed and stagnation, because of the mounting attempts to raise socio-economic and health standards under the driving force of independence and foreign aid.

Not only the under-developed world is concerned in these changes. The highly sophisticated worlds of Europe and North America, of South Africa and Australia are changing, too.

The geography of medicine has changed completely since the jet made its appearance and the changes have been in many respects too swift for the geographer and the doctor alike.

Two major factors are involved in this, namely, mass immigration of labour from the poorer parts of the world and the continued and increasing movements of individual travellers (Maegraith 1963).

## THE IMMIGRANT

Considerable mass migrations are taking place from developing countries to the wealthy lands of Western Europe, the United Kingdom and North America. These are dictated largely by the need for labour in the latter, arising as a consequence of the high socio-economic and living standards and the progress of the sophisticated countries.

It has been estimated that in the last five years more than six million immigrants have entered Europe and the United Kingdom and very large numbers have gone into the U.S.A. despite the legal limitations put on immigration.

The political and economic significance of this influx of foreign labour in respect of both recipient and donor countries is well recognized, but the medical implications have only very recently been seriously considered. The standing Congresses of Schools of Tropical Medicine of Europe noted in 1965 that these gigantic movements of labour were creating in Europe a completely new medical picture which was of basic significance to the public health and well-being of the sophisticated countries and to the doctors practising in them. The migrations are in fact once again changing the medical geography of the developed world.

What this means in terms of a single country was illustrated recently by Hofman (1964) in reference to Switzerland, the indigenous population of which is only about five and a half million. Almost the same number of tourists visit the country each year. Officially registered foreign manual labour in 1964 varied from season to season between 500,000 and 700,000. These workers came from Southern and Eastern Europe, Turkey, the Middle East, North Africa and some from black Africa and even the Far East. Some of them, including North Africans, were domiciled in France and worked in Switzerland, so escaping the usual medical scrutiny at the borders. The numbers of immigrants are now fairly strictly controlled, but the figure in 1966 was reckoned at nearly a million.

The immigrants bring their own diseases with them. The nearly 300,000 Swiss who live abroad (more than 50,000 in tropical countries) regularly return to their native land and sometimes bring back with them infections acquired in their exile. Hundreds of thousands of Swiss take holidays each year in distant countries, many in the eastern Mediterranean littoral, the East and Far East and in tropical Africa and South America. Some of them also return with exotic infections.

Switzerland thus represents a little world on its own, reflecting the larger global picture of changing medical patterns resulting from introduction of exotic disease by travellers, by visitors and by immigrants. To this may be added the significant point that Switzerland imports the greater part of her food commodities, including live animals and carcases, so that the serious possibility exists of introduction of animal diseases which could involve man, including worm infections, rabies and other viral infections.

It is no wonder, therefore, that Switzerland now has in Basel its own Institute and Hospital for Tropical Diseases.

Aujoulat (1964) has recently analysed the immigration of Africans into France. Apart from the politically special group of Algerians, who numbered some 500,000, black African migrants now living in France amount to about 50,000. In the past, the latter came mostly from the coastal belts and big towns of West, East and Central Africa, and the individuals were usually socially well advanced, with some professional or skilled training, a good knowledge of French and were physically fit. Most of them entered metropolitan France under supervision for further education and training.

In recent years a new migration of Africans has been taking place composed of much less advanced individuals with practically no knowledge of industrial or artisan life and financially and physically poor. These migrants have been largely Sarakollé people from Mali, who come from one of the poorest countries in the world, where the land provides a miserable output of millet, peanuts, maize and cassava inadequate to support the increasing population. Their primitive local trade is conducted by barter, mostly of cereals for livestock.

As with many other poor agricultural communities, seasonal emigration to the better conditions of neighbouring countries has been the rule for centuries. Until a few years ago the main cyclical movement of labour was into Senegal for the ground nut industry. There was also temporary exodus to the coastal plantations of Guinea and as far south as the Congo, and some recruitment in the French Army and Merchant Navy.

Since the independence of Mali these traditional currents of migration have slowed almost to a stop. Contiguous countries are now also independent, and the consequent balkanization has grossly restricted local migration, as each country attempts to reserve to itself the exploitation of its natural resources. So the Malis, no longer able to move into Senegal or the Sudan, have turned instead towards their old colonial power, metropolitan France.

The immigrants from Mali are almost exclusively males between twenty and forty years of age. They are almost all agricultural peasants who form a small force of employable manual labourers; there are few with even artisan skills. In France they have congregated mostly in the Paris area and have regrouped ethnically into 'villages' with elected Chiefs and Elders, where they can lead a life as near as possible to their own in Africa. Their impact on metropolitan France is in itself slight. They represent a temporary phenomenon which could soon be controlled by agreement with the African States concerned. Nevertheless, such migratory currents from the tropics are influenced by economic factors which are inherent in under-developed countries.

The migration of Sarakollé into France is unusual in that political closure of the normal internal routes of migration has forced them to leave the continent of Africa. Otherwise, it represents the common movement which occurs in any country where a basically rural population is trying to reach industrial and artisan status and which is seen not only in emerging countries, but also in many parts of Western Europe and the United Kingdom and Ireland, wherever the industrial twentieth century calls.

Politically, this sort of migration, even the recent new influx of Algerians and their families into France, may be helpful to both sides, if the metropolitan countries, which receive the immigrants, however unwillingly, will offer them a chance to learn a trade or a profession useful to their homelands when they return. This might be the best form of co-operation, helping the emerging countries in their evolution. In turn, as these countries developed, migration from them to the sophisticated world would find its own limits.

This may well be so, but in the meantime, the pressure of migration from poor conditions to the wealthier parts of the world is increasing, and bringing with it immediate problems of health. Disease introduced or re-introduced into an area by immigration may primarily affect the immigrants themselves, the inhabitants of the recipient country, or both. Whether the nationals become involved depends largely on the nature of the introduced infection, the degree of mixing of immigrants and indigenes and the possibility of the survival of the infection in the local environment.

Bacillary dysentery or virus infections such as smallpox which can be directly transmitted from man to man could spread anywhere in a non-immune population. The introduction of infections which require intermediate hosts such as mosquitoes or snails before they can be transmitted does not put the local community to risk unless the intermediate hosts are available and the environment is suitable.

In some areas, as in southern Spain and Italy, the latter conditions do exist for specific infections which may consequently become implanted or re-introduced. This is the case in the 'Midi méditerranéen' in France where the meteorological conditions, and to some extent the fauna and flora, resemble those parts of North Africa and the Middle East (Harant and Rioux 1964). Amongst the diseases which have become established in this region are virus fevers, rickettsial infections including fièvre boutonneuse and visceral (kala azar) and skin leishmaniasis. The authors note that at the moment hookworm infection has disappeared from the area, but consider that mass importation of African workmen might induce its reappearance. Perhaps it is just as well in the latter respect that the climate of England is as miserable as it is, since it has been demonstrated recently that it is possible for hookworm to be transmitted in the Midlands.

## THE TRAVELLER AND DISEASE. UNDE VENIS?

The health barriers which in the past have protected Western Europe and North America from serious imported infectious disease are now coming under enormous pressure not only from immigration but from overseas visitors and travellers returning home from abroad.

The dominant factor is the commercial aeroplane which has led to a remarkable increase in speed of travel over the last two decades.

The relatively low cost of flying has, moreover, greatly increased the numbers of people travelling. Many who seldom left their homes in sophisticated countries now travel through the world as a matter of course on vacation or business, often to the poorer countries of Africa and the East, where the medical situation is primitive in comparison with that in their home country. The reverse also occurs but somewhat less frequently and in smaller numbers.

Either way, there is increased movement to and from the tropics of politicians, businessmen, missionaries, professionals, air crew and servicemen, holiday makers and those going to assist developing countries, architects, engineers, experts on financial aid. School children from Europe visit their parents in the tropics on vacation. This shuttling is done at such speed that a man can return to his home country from across the world in a matter of hours, well inside the incubation period of any known infectious disease.

Ships in the past took so long to go from an endemic foreign area to the home port that infections acquired abroad usually declared themselves before the journey was completed and disease in travellers and crew alike was visible and easy to detect by the efficient public health authorities.

Today, however, the air traveller may present himself to the health authorities apparently perfectly well, but with a disease upon him which will not become overt until days or even weeks later.

Time has become a vital factor in the control of disease and its distribution. It is not surprising therefore that the health barriers and regulations in ports and airports are unable adequately to cope with the new conditions.

In this respect, the breakdown has occurred not only at the public health barriers. There has also been a serious failure of doctors in the sophisticated countries to detect, diagnose and treat such exotic infections when they declare themselves either in immigrants or in the returned traveller. This means that outbreaks of infectious diseases such as smallpox, and fatal accidents from undiagnosed falciparum malaria, for instance, may be expected, and indeed, are from time to time occurring.

The continuing introduction of exotic conditions into the sophisticated world and the failure of the usual health barriers inherent in the speed with which the traveller may move inside the incubation period of an infectious disease has thus put the medical practitioner in the first line of defence in public health. In some respects this is unfortunate because his training and outlook do not yet suit him for the task. He has not been trained to think of the possibility that the next patient he sees might have malaria from the West African coast, typhoid from Iraq or may be in the early non-specific stages of smallpox acquired in India, Indonesia or Brazil. He has not realized that the geographical movements of a patient are now an integral part of his medical history. He has not been taught to ask the simple and vital question of the traveller—*Where have you been and when?* and of the immigrant —*Where have you come from and when?* (Maegraith 1965).

Even if the practitioner asked these questions, he would be, more often than not, unable to interpret the answers, because he is practically always essentially ignorant of the global distribution of the more serious and commoner infectious diseases.

If the practitioner in Britain or his equivalent in Europe and America or Australia is to do his job efficiently in this jet age he must be orientated towards the search for and diagnosis of imported disease. Medicine has become global rather than local. The doctor can no longer regard himself as working exclusively in Liverpool or Adelaide. He has to accept that he is practising in a shrinking world, and that the disease from outside may affect him and his patients at any time.

The doctor of today and tomorrow must be tuned to global medicine, and for this the present profession at all levels needs re-education and must have some knowledge of the world distribution of the more important infectious diseases.

This is not the place to discuss how this should be done but it is necessary to point out that what has been said for the doctor in the sophisticated world applies in many respects to his equivalent in the developing countries themselves. Not only may disease from outside be introduced into the latter by travellers and planners and consultants, but redistribution of disease patterns inside a given country will take place wherever socio-economic developments and the corresponding movements of labour and resettlements of populations occur. The same is true of war, as may be seen today in South Vietnam.

The practising doctor in the emergent nation should be aware of this and should moreover, have at least a working knowledge of local endemic diseases and their distribution. To ensure this for the future will require reorientation of the local medical student curriculum towards endemic and community medicine. Only in this way will the supply be ensured of doctors capable of tackling the problems of the infections which hold back the rural masses of the tropics.

## THE POPULATION EXPLOSION

The growing medical control of infectious disease in the developing countries, with consequent reduction in morbidity and mortality has already led to remarkable and in some areas almost overwhelming population increase in the last twenty years. If left unchecked, this will have a profoundly depressing effect on the standards of living of untold millions and will directly and indirectly alter the medical geography of the world.

Further control of disease and malnutrition might be expected to augment the problem by upsetting the precarious balance of terrestrial life by stimulating ever greater rates of population growth in just those areas where living conditions are already bad and food supply and distribution minimal.

Sabin (1962) has stated the proposition as follows:

> . . . there are many who believe that the continued acquisition and application of knowledge designed to prolong life and to eliminate the miseries and handicaps of disease, without comparable activities directed at fertility control, will in effect, before long, result in the even greater miseries of hunger, poverty and war, and thereby create the greatest potential threat to human survival in all parts of the world.

Current methods of fertility control are limited in application and are still unsuitable for the hundreds of millions of economically depressed peoples who need some means of family limitation. Research into this problem is urgently needed, as it can fairly be regarded as potentially the most important public health problem in the world today. At the moment it is a factor that affects the rural masses of the tropics unevenly. India is already in trouble with over-population. On the other hand, Ghana is under-populated and economically and agriculturally capable of supporting more inhabitants. To this extent the problem may sometimes appear largely local and political, but in the long run it is global.

Population pressure arising from disease control emphasizes the fact that no one great social movement is adequate by itself. As doctors, we may change and are changing the medical maps of the world, but the desired result, the betterment of the health and living standards of mankind, will never be achieved unless education, agriculture and the distribution of food, and veterinary science move forward with us.

If nothing is done, the grim warning quoted above will become reality.

## BIBLIOGRAPHY

AUJOULAT, L. P. (1964). 'Aspects statistiques et demographiques de la main-d'oeuvre importée des pays chauds et tropicaux'. *Bull. Soc. Path. Exot. Paris,* **62**: 716.

CANDAU, M. G. (1966). 'Mary Kingsley and the work of the Liverpool School of Tropical Medicine'. *Ann. Trop. Med. Parasit.,* **60**: 1.

CIBA FOUNDATION (1965). *Man and Africa.* Churchill, London.

Conference des Ecoles et Instituts Européens de Médecine Tropicale et d'Hygiene (1963, 1964). Rénnions 2 et 3. Inst. Med. Trop. Prince Léopold, Anvers.

DAVEY, T. H. AND LIGHTBODY, W. P. H. (1961). *The control of disease in the tropics.* H. K. Lewis, London.

HARANT, H. AND RIOUX, J. A. (1964). 'Etat actuel de quelques maladies des pays chauds importées en Lanquedoc–Roussillon', *Bull. Soc. Path. Exot. Paris,* **62**: 867.

HENSCHEN, F. (1966). *Gründzuge einer historischen und geographischen Pathologie.* Springer-Verlag, Berlin.

HOFMAN, E. (1964). 'Maladies exotiques importées en Suisse', *Bull. Soc. Path. Exot. Paris,* **62**: 781.

MAEGRAITH, B. G. (1963). 'Unde Venis?' *Lancet.*

MAEGRAITH, B. G. (1965). *Exotic Disease in Practice.* Heinemann, London.

MAY, JACQUES, M. (1958-61). *Studies in Medical Geography.* American Geographical Society, New York. 3 volumes.

PRICE, A. G. (1949). *White Settlers and Native Peoples.* Georgian House, Melbourne.

PRICE, A. G. (1963). *The Western Invasions of the Pacific and its Continents.* Clarendon Press, Oxford.

PROTHERO, R. MANSELL (1965). *Migrants and Malaria.* Longmans, London.

SABIN, A. B. (1962). *Preface to Tropical Health.* National Academy of Sciences, National Research Council, Washington D.C.

SOPER, F. L. AND WILSON, D. B. (1943). *Anopheles gambiae in Brazil, 1930-1940.* The Rockefeller Foundation, New York.

STAMP, L. DUDLEY (1964). *Some Aspects of Medical Geography.* Oxford University Press, Oxford.

STAMP, L. DUDLEY (1964). *The Geography of Life and Death.* Collins, London.

DAVID LEA

# *Some Non-nutritive Functions of Food in New Guinea*

*'The shortest residence in a tropical country and the briefest examination of tropical literature bring home to the student the importance of diet.'*

A. GRENFELL PRICE *White Settlers in the Tropics*

IN STUDIES of agrarian geography among subsistence farmers, geographers have tended to emphasize production and transportation to the almost complete exclusion of distribution and diet.[1] Diet, however, expresses a relationship between man and his environment just as does production, for ecological conditions can determine the chemical composition and abundance of foods, and climate, for example, can set nutritional requirements (Sorre 1962: 449). Conversely food preferences can result in transformed environments, and methods of food preparation and cooking can result in deficiency diseases. Hunger or malnutrition in particular are ecological phenomena for they are 'a manifestation of disequilibrium between human groups and their social and physical milieu' (Sorre 1962: 453). However, apart from studies of the physiological purposes of food and the dramatic problems of constant, periodic or episodic hunger (Clements 1967) many other studies are of interest to geographers.[2] For example, diet can reflect the cumulative weight of tradition, beliefs, taboos and prejudices, while problems connected with waste, storage, methods of distribution of food, nutritional stresses, times and areas of food shortage, are just as important as studies of food production itself.

In a previous publication (Lea 1965: 201-4) the author made *inter alia* a comparative study of diet in two Abelam villages in the Maprik area of New Guinea. In that article two points were made

[1] For studies on a world scale see de Castro (1952) and Kariel (1966). Kariel's article also contains references to a few detailed dietary studies by geographers. Needless to say there are numerous studies by nutritionists but their terms of reference and interests are different from those of a geographer.

[2] For example May (1961, 1963), Simoons (1961), Bennett (1962).

which are relevant to this present study: first it appeared that apparent food surpluses existed even when there was patent under-nutrition; secondly it was suggested that where there was insufficient food and poor diet, not only were the health and productive capacity of the people impaired, but the social and ceremonial life of the people suffered. Of these two points the first is most significant. Where quantitative studies of both production and consumption have been made among groups using a bush-fallow system, huge surpluses appear to be usual.[3] For example, the Bureau of Statistics of Papua and New Guinea (1963: 15) estimates that throughout New Guinea eleven pounds of food from the principal crops and sago are produced per person per diem. If five pounds are allowed to satisfy daily physiological needs (Hipsley and Kirk 1965: 77; Bailey 1963: 6) and say half a pound is reserved for planting material, about half the total production represents an apparent surplus. While errors in sampling, weighing and estimation of consumption may account for some of it, it is inconceivable that at least a large surplus is not produced. What happens to this surplus and why is it produced?

The main reason, no doubt, is that subsistence societies aim to produce a surplus to ensure against losses by flood, drought, wild animals, insects, theft and losses in storage. This is what Allan calls the 'normal surplus' of subsistence agriculture (Allan 1965: 38). If a surplus still exists after all dietary needs are satisfied, surpluses can be marketed, stored or given to animals and thereby can be converted into wealth or a vegetable or meat reserve. Also in the manner of the potlatching practices of the Indians of the north-west coast of North America, they may be wasted or destroyed for reasons of prestige (Vayda 1961).[4]

This article is an attempt to illustrate in some detail a further need for a surplus. It is suggested that as food is an integral part of payments, distributions and exchanges, a surplus is essential for the proper functioning of society. Although large quantities are consumed at feasts, these activities account mainly for periodic or occasional surpluses—not permanent surpluses—for all of these functions are usually reciprocal and food distribution takes place on an inter-community basis which tends to even out local, seasonal or annual inequalities in the possession of food (cf. Vayda 1961).

[3] Some examples are cited in Lea (1965: 203).

[4] It is not unusual for the Abelam to become destructive after an argument. Yams often are destroyed in a garden after a man has had a dispute with his wife.

## NORMAL CONSUMPTION

The Abelam family as a unit produces nearly all its food in gardens, using techniques typical of Melanesian lowland shifting cultivation.[5] The main crops are yams, taro and bananas, supplemented during seasonal shortages by sago. The family likewise consumes the bulk of the food although it is rare for the single family to have a meal without at least some temporary accretions or depletions.

The main meal of the day, often a thick yam soup, is the evening meal prepared by the women of the household: leftovers from this meal are usually eaten as a quick snack in the morning. During the day, while all but the old and sick are out working sago or in the gardens, the women prepare a small meal wherever work is being done. This meal usually consists of baked, boiled or steamed yams, taros or sweet potatoes. Often in the course of a day both men and women eat coconuts, pawpaws, sugar cane (if a taboo does not apply)[6] and sweet bananas or, if a small snack is required, they prepare breadfruit, 'pit pit' (the inflorescence of the *Saccharum edule*) or bananas which can be quickly cooked over an open fire. These are, however, incidental foods and the four staple foods, yams, taro, bananas and sago, account for 85-95 per cent of calorific intake.

Most evening meals are eaten near the family hearth but if a husband is visiting in his own or a contiguous hamlet, his wife will take a bowl of soup or some cooked yams or taro over to him and he will then share it with the men who are with him. Sharing of food is inculcated from an early age and any friend or kinsman is invited to partake if he happens to be around when food is being eaten.

In the following sections this theme of social functions of food will be developed and some aspects of Abelam life will be described which will illustrate that food is a part-payment for services and, further, how it is essential to provide for the proper functioning of the ritual and ceremonial life of the people.

[5] For an account of these techniques see Barrau (1958) and Watters (1960). Among the Abelam some food is grown in special ceremonial yam gardens, some is obtained by hunting and gathering and from planted tree and sago groves, and some is bought from trade stores (Lea 1964: 123-9).

[6] For a discussion of Abelam food taboos see Kaberry (1941: 364-5). Some effects of these food taboos are discussed in Lea (1964: 137). For an interesting discussion of the distribution and causes of meat taboos in the Old World see Simoons (1961).

## PAYMENTS FOR SERVICES

In many societies, especially those without a money economy, gifts of food can serve as a part-payment for work and services. De Schlippe in describing the working beer parties of the Zande saw them as providing incentive and rewards for services rendered (de Schlippe 1956: 148). In the same way among the Abelam a woman who helps one of her kindred extract sago flour is usually given part of the flour for her own use, or when a man attracts a large workforce to help him plant his yam garden, he usually gives the workers yam soup and a small gift of native tobacco, betel nut and pepper vine catkins (which are chewed with the betel nut). Again in the common task of house building[7] the future owner of the house assembles fifteen to twenty-five helpers to thatch the house. Around mid-day the wife of the house owner gives the men a simple meal of steamed or baked yams, cooked bananas or taros. When the work is finished in the late afternoon, the wife provides a final meal of the highly esteemed white yam soup.

Among the Abelam it must be realized that in all cases (including the two described in greater detail below) help is often given on a reciprocal basis and it is extremely difficult to separate 'hospitality' from 'payment': certainly any payment of food is never complete for, as Kaberry (1941: 86) says, 'immediate equivalence is never obtained for there is a chain of rights and obligations extending through the whole life of the individual'.

### *Building Tambaran Houses*

When a tambaran or cult house is built,[8] payments in food are similar to those examples already cited although the amount of food given is greater and payment is made with greater ceremony. Surrounding allied villages, but even some traditionally hostile groups, send some form of help. Some villages send skilled artists for painting the façade, others building materials, and some labourers who help with big tasks such as the thatching. When the work is going on or when the delivery of building materials is made, all visitors are given soup and large quantities of cooked yams; sometimes portions of uncooked pig meat are given to the visitors to take home. When the tambaran house is finished there is a large ceremonial distribution organized by the men of the *amei*

[7] For details of the techniques of house building see Kaberry (1941: 84).

[8] The tambaran houses are huge tetrahedral shaped houses, up to 70 feet high, used for ceremonies and storing the men's sacra. There are only two or three such houses in each village.

concerned. One such occasion at Yenigo village in 1963, provided an opportunity to record the amounts of food distributed.

> Placed in a line across the middle of the *amei*[9] were 35 bowls of white yam soup each containing about twelve pints of soup. Accompanying each bowl and placed in a leaf sheath of the 'limbom' palm, was approximately 40 pounds of yam . . . two coconuts, some native tobacco, some betel nut, some leaves and inflorescences of the *Piper betel* and an uncooked piece of pig meat—in all 35 lots of food. . . . It seemed that food was given to each hamlet that helped. The cooked food was immediately consumed and the other produce was divided among the helpers from the various hamlets.

These distributions need considerable preparation, foresight and planning and involve about half the people of the village. This particular distribution alone required almost one ton of yams which is about the same amount eaten by a whole family in a year. Distributions of this sort definitely account for the planting of extra land with yams in the years that they take place and they are postponed if the crops fail.

### *Initiations and Other Feasts*

All men within the village belong to one of two unnamed, non-exogamous and non-localized sections called *ara*, which are usually but not rigidly patrilineal. Every adult male has an exchange partner called *tshambera* in the opposite *ara* and this relationship is significant during initiations and the annual exchange of large yams known as *wabi*[10] (Lea 1966). During the initiations the young male initiates are segregated from the rest of the village in an enclosure near the tambaran house.[11] One *ara* of the village acts as initiators for the other *ara* and feeds the initiates with white yam soup, all other foods being taboo. The soup is eaten out of half coconut shells that are used only once. When all the soup is eaten the plates are put to one side and are later threaded on a cane rope and hung in a prominent place before the tambaran house. It is the aim of the initiators to make the initiates as fat as possible and to be able to boast of the large quantities of food consumed.[12] The

[9] The *amei* is a flat cleared area in the middle of the main hamlets of the village. It is the site of most activities, ceremonial, economic and social.

[10] These *wabi* are usually *Dioscorea alata* but certain varieties of *D. nummularia* and *D. esculenta* are sometimes included.

[11] Initiates may be between twelve and twenty-four years of age and occasionally younger.

[12] Rather ambivalently the women are told that the initiates have eaten nothing but what they were able to find in the bush (Neve 1960: 123).

initiating *ara* receive a presentation of pigs from the other *ara* after the initiations have taken place. Each *ara* takes it in turn to act as initiators.

During the initiations there are many visitors to be fed and there are a number of associated feasts. The initiations require such vast quantities of food and wealth to buy pigs that a gap of several years is necessary before any further ceremonies can be undertaken. Forge (personal communication) estimates that the minimum time for each full initiation cycle is about ten years and he thinks that it may be even longer. The feasts bolster the prestige of a village and may establish or reinforce ties between groups and individuals and will make individuals clearly state their allegiance to certain groups. The timing of each individual ceremony is not fixed by an astronomical, ceremonial or ritual calendar but they are always held after the yam harvest and usually at a full moon. They can, if necessary, be delayed until adequate surpluses of food can be built up. This applies to the building of tambaran houses, to ceremonial exchanges, and even to marriages. Feasts that must, by tradition, be held when the events take place, such as births, deaths and girls' puberty, are usually less elaborate and can take place even in times of food shortages.

## FOOD IN EXCHANGES

The distinction between payments and distributions of food on the one hand and exchanges on the other is somewhat arbitrary for both require equivalence and reciprocity in some form. Exchanges can be defined here as events at which careful tallies are kept, and where the stated object is that equivalence with the object exchanged should be obtained as soon as possible. However, 'exchanges' are also the occasion of 'distributions' and the following description of the ceremonial yam or *wabi* exchange illustrates the mechanics of the most important of the exchanges.

### *The Wabi Exchange*

The preparations for a yam exchange and all the associated ceremonies take about a week. The men carefully decorate the tubers which are 3 to 10 feet long and both the men and the women collect food, firewood, betel nut and leaves of the pepper vine; new skins have to be put on some of the drums and there is general excitement. During the night before the parade of the yams, there is dancing and singing lasting until dawn. Just after dawn all the big *wabi* are carried on their poles into the centre of the *amei* and are put in lines under shelters made of coconut fronds. The *wabi* are extensively decorated with both basketry and wooden masks, shell rings, feathers, leaves, berries and many ceremonial trappings; some of them are made to look remarkably like human figures.

The yams are admired, compared and criticized by men from the home village and from other villages. There is much formal talk where a man, shaking spears in an aggressive fashion, struts to and fro across the *amei* chanting his speech. Most of the men parading yams and a few of the visitors also do this and the chants are mainly bombastically abusive or self-congratulatory. Some chants are about the spirits involved in yam growing, the ancestors and their yam growing activities; some are mainly admonitory and exhortative and directed at the young men to encourage them to take up yam growing and to stop being so interested in women.[13] A visitor often congratulates the village publicly but makes maledictory comments around the *amei*.

During the day the yams are displayed and the visitors fed. The decorated yams remain on the *amei* overnight and the men sit up talking and watching over them. The following morning the exchanges take place and all the yams displayed are given to the *tshambera*, the ceremonial exchange partner in the opposite *ara*, who may later give the yam to one of his kindred or use it later himself for food, planting or in some hostile exchange. After the exchange the decorations are removed and returned to the original owners and the yams are taken to the yam house of the new owner, stored in a tambaran house or hung in the rafters of one of the 'talking houses' which contain the slit gongs.[14] Often the largest of the yams is left on show until it rots and if an exchange is rejected or there is some dispute the yams concerned are left in a prominent place to rot. The implication here is that 'you are afraid to accept my yams because you cannot return similar samples. I have yams to spare. I let mine rot.'

The day after the exchanges are made, some of the men from both *aras* in the village set out and buy pigs with shell rings and money from other villages. In one exchange observed twelve pigs

[13] Women are held to be inimical to the growing of ceremonial yams and the strongest of all taboos is on sexual intercourse while the yams are growing.

[14] Slit gongs are logs hollowed out from only a narrow slit along one side of the trunk. Both ends of the log are solid. Hit on the side with a piece of wood simple messages, call signs and dance rhythms can be played upon them.

were exchanged and this was about one-quarter of the pigs in the village. When all the pigs are assembled in the village, three or four days after the yam exchange, they are each lashed on to a single pole and placed alive over a fire which burns the hair off the pig but does not always kill it. The pigs are then cut up and all the portions of raw meat are given to *tshambera*. Groups of men from one *ara* give the pig to their *tshambera* in the opposite *ara* who may give small portions to visitors from other villages. These gifts are either reciprocating a similar gift already made or creating an obligation which will ensure equal hospitality in later exchanges in other villages. The final exchange associated with the *wabi* exchange is when each *ara* prepares and gives yam soup to the members of the other *ara*.[15] At this time some *ka* (small non-ceremonial yams mainly *Dioscorea esculenta*) are also exchanged and each *ara* presents a large conical heap of *ka* to the other *ara*. All these exchanges associated with the original *wabi* exchange may last for about a fortnight and nearly every night of that fortnight there is dancing and singing throughout the night.

*Hostile Exchanges*

Hostile exchanges are an institutionalized form of hostility. If two men, two groups, or even two villages have a dispute, hostile exchanges of yams and pigs usually take place. Although the proximate causes of dispute are often trivial they usually have very deep-seated origins and exchanges often reach gigantic proportions. One hostile exchange started because a man spat in the face of another man from a traditionally hostile village. It grew in size as each village strove to outdo the other and ultimately involved a string of allied villages and two and a half tons of yams and a number of pigs. Allies of the receiving village were given some of the food in payment for past help and in anticipation of future help.

When exchanges are made by a group, yam soup, tobacco and betel nut are given to the visitors who receive the presentation. When the presentation is made there is usually much disputing, shouting and mocking by both parties. Bombastic and inflammatory speeches are made about the nature of the gift or about the manner in which it is accepted. If the receiving party refused to accept yams the implication would be that they were unable to grow similar or bigger yams and to refuse pigs would imply

[15] In Yenigo in one exchange 46 bowls of soup (about 60 gallons) changed hands.

poverty in shell rings with which pigs are normally bought. These exchanges of institutionalized rivalry serve as a controlled expression of both permanent hostility and temporary disputes which could otherwise lead to more harmful conflict and the disruption of society. The exchange relationship can 'only be terminated when passions cool and equivalence is obtained' (Forge: pers. comm.) and there is a ceremonial exchange of rings and sometimes a ceremonial planting of coconuts.

### *Marriage Exchanges*

An interesting example where payment and exchange are inseparable is marriage. However, marriages differ from other Abelam exchange relationships in that exact equivalence is not expected. The wife givers give the woman, yams (often well over a ton) and later, ceremonial services. In return they receive from the husband's group highly prized shell rings as the bride price again, at the birth of the wife's first child, for ceremonial services, and perhaps seven decades later the wife givers' heirs receive shell rings on the deaths of all the wife's children.

## CONCLUSION

Food therefore has more than a direct physiological function in subsistence societies. The Abelam, for example, certainly aim to produce enough to satisfy dietary needs and to provide a 'normal surplus' against natural disasters. There are a number of ways in which this 'normal surplus' can be disposed of but it is variable and is made deliberately greater in some years so that production —and subsequently consumption—varies due to social and ceremonial needs. These needs may be to make a payment for goods or services; to feed pigs to build up wealth for special functions or to buy highly prized shell rings; to provide a medium of exchange and thus a means of acquiring prestige and of creating and meeting obligations. Food, and therefore the surplus, also has an aesthetic value, for the Abelam are a hospitable people and they 'take a keen delight in contemplating large quantities of food' (Kaberry 1941: 353). Reay describing the Kuma of the Central Highlands of New Guinea also shows how food is highly esteemed for its symbolic function (Reay 1959: 89-95). *Mutatis mutandis* these conclusions will apply to any society but diet, prejudices, other uses of the surplus, such as green manures or burnt offerings—either to the gods or to maintain prices—could well explain apparently irrational decisions or particular distribution patterns.

Much more work still needs to be done on the disposal of surpluses. In the case of exchanges, particularly hostile exchanges, it is possible that an exchange is initiated to dispose of an immediate surplus of fresh food in the hope of receiving a back payment later when stocks of food are exhausted.[16] This could well be important among peoples growing vegetable crops which cannot be preserved for long periods. Numerous other questions arise concerning the relationships between production and consumption. What actual changes are made in land use in ceremonially active periods? How do these changes affect labour and general productivity within the village? What are the inter- and intra-village movements in pigs and yams? To what extent does crop failure affect an already planned function and to what extent does time of harvesting affect timing of activities? Answers to these questions will only be found by detailed studies in a local community over many years.

[16] This is similar to the exchange of *pandanus* nuts in the Central Highlands of New Guinea (Hipsley and Kirk 1965: 56).

## BIBLIOGRAPHY

ALLAN, W. (1965). *The African Husbandman*. Edinburgh.

BAILEY, K. V. (1963). 'Nutrition in New Guinea', *Food and Nutrition Notes and Reviews,* **20**: 3-26.

BARRAU, J. (1958). *Subsistence agriculture in Melanesia*. Honolulu.

BENNETT, C. F. (1962). 'Bayano Cuna Indians, Panama: An ecological study of livelihood and diet', *Annals Assoc. Amer. Geogr.*, **52**: 32-50.

BUREAU OF STATISTICS, PAPUA (1963). *Survey of Indigenous Agriculture and Ancillary Surveys*. Konedobu.

DE CASTRO, J. (1952). *The geography of hunger*. Boston.

CLEMENTS, F. W. (1967). 'The geography of hunger'. *Aust. J. Sci.,* **29, 7**: 206-13.

HIPSLEY, E. H. AND KIRK, N. E. (1965). *Study of dietary intake and the expenditure of energy by New Guineans,* South Pacific Commission Technical Paper No. 147.

KABERRY, P. M. (1941). 'The Abelam Tribe, Sepik District, New Guinea. A preliminary report', *Oceania,* **11**: 233-58, 345-67.

KARIEL, H. G. (1966). 'A proposed classification of diet', *Annals Assoc. Amer. Geogr.,* **56**: 68-79.

LEA, D. A. M. (1964). Abelam land and sustenance. Unpublished Ph.D. Thesis, Australian National University, Canberra.

LEA, D. A. M. (1965). 'The Abelam: A study in local differentiation', *Pacif. Viewpt.,* **6**: 191-214.

LEA, D. A. M. (1966). 'Yam growing in the Maprik Area', *Papua & New Guinea Agricult. J.,* **18**: 5-16.

MAY, J. (1961). *The ecology of malnutrition in the Far and Near East*. New York.

MAY, J. (1963). *The ecology of malnutrition in five countries of Eastern and Central Europe*. New York.

NEVE, J. (1960). 'Spirit houses of Maprik', *J. Public Service T.P.N.G.,* **2**: 118-25.

PRICE, A. G. (1939). *White Settlers in the Tropics*. American Geographical Society Special Publication No. 23, New York.

REAY, M. (1959). *The Kuma*. Melbourne.

DE SCHLIPPE, P. (1956). *Shifting cultivation in Africa: The Zande System of Agriculture*. London.

SIMOONS, F. (1961). *Eat not this flesh*. Madison.

SORRE, M. (1962). 'The geography of diet' in *Readings in Cultural Geography*, ed. by P. L. Wagner and M. W. Mikesell from 'La geographie de l'alimentation', *Annales de Geographie,* **61** (1952): 184-99.

VAYDA, A. P. (1961). 'A re-examination of Northwest Coast economic systems', *Trans. New York Acad. Sci.*, Ser. II, **32**: 618-24.

WATTERS, R. F. (1960). Some forms of shifting cultivation in the South-West Pacific. *Jour. Trop. Geog.*, **14**: 35-50.

DIANA HOWLETT

# *Australia in New Guinea: None So Blind . . .*

*'The vast, rugged and tropical Island of New Guinea . . . presents one of the leading and most explosive problems of current geopolitics.'*

A. GRENFELL PRICE *The Western Invasions of the Pacific and its Continents*

## INTRODUCTION

COLONIALISM and under-development—two historical phenomena of transcendent importance—are now extensively documented and widely understood. For good reason: the experience with both is long, the record voluminous. A literature on bibliography, itself, is beginning to emerge (Spitz and Weidner 1963; Jones 1964; Bicker 1965; Mezirow 1963). The two-thirds of the world's area and people which in the last century and a half were or are colonies, were and are under-developed. This is not to argue the reductionist claim that colonialism alone caused under-development. But the correlation is positive and significant; and the interaction of their major aspects, plain.

A great store of comparative knowledge has therefore accumulated. That the economic, social, demographic, political, and other salient characteristics of backwardness have been, and may be, affected by colonial policy is demonstrable. With insight and certainty, colonial powers may regulate many of the strains of independence and of subsequent national development.

In this period of enlightenment, Australia remains one of the few colonial powers. It is the purpose of this essay to examine her policy in one of the last colonies, New Guinea.

## THE SETTING

As it is currently administered by Australia, New Guinea comprises the Australian possession of Papua in the south-east sector of the island of New Guinea, and in the north-east the former German colony of Kaiser Wilhelmsland. Australia was given a mandate over the German colony after the First World War; after the Second World War its status was changed to that of a United

Nations Trust Territory under Australian administration. Australian jurisdiction also extends over several hundred adjacent smaller islands to the north and east of New Guinea.

Papua and the Trust Territory are roughly equal in size, with a combined area of about 183,000 square miles. A massive central cordillera extends east-west through the country, effectively isolating the northern and southern lowlands. Transport and communication between the north and south is by air or coastal shipping. There are no north-south roads, no railways anywhere in the Territory, and few rivers are navigable for very far inland by craft larger than canoes. Not only are conditions of terrain and climate here among the most extreme, making difficult and expensive both the establishment and maintenance of an economic infrastructure, but, in common with many tropical lands, New Guinea has virtually no economically useful minerals or sources of energy. The gold deposits which produced the bulk of pre-war revenue were exhausted by the mid-1950s; oil exploration has been extensive but so far unrewarded; extensive reconnaissance by Australia's Bureau of Mineral Resources has not located coal or iron. Some known deposits of metals, such as nickel, are considered too costly to exploit. Similarly, supplies of natural gas are located in western Papua's swampbound terrain, hundreds of miles from urban centres. Copper deposits on the island of Bougainville, as yet unworked, hold the only promise for a mining industry at the present time.

In 1966, the indigenous population was just over two million, and the non-indigenous population almost 35,000. The term 'non-indigenous' includes all those other than the ethnically indigenous people, whether or not they were born in the two Territories. Thus the term embraces Europeans of many nationalities but mainly Australian citizens and Asians, and mixed-race people. The average population density in Papua is about 7 persons to the square mile, and in the Trust Territory 17 per square mile, with an overall average of 12 persons per square mile. The rugged and inhospitable terrain over much of New Guinea, however, means that these crude figures of population density conceal considerably greater actual densities in some regions.

The population is unevenly distributed. About 40 per cent of the people occupy a series of broad highland valleys lying at an altitude of 5,000 feet, and their steep mountain flanks, in the central cordillera. Local population densities here may be as much as 500-600 per square mile. Other heavily populated districts are the Gazelle Peninsula at the northern end of New Britain, and the Sepik Valley in the north-west of the Trust Territory. Most of the coastlands are moderately populated, with densities ranging from 20 to 100 per square mile. The zone between 2,500 feet and 4,500 feet is virtually unpeopled, being steep, dissected, heavily forested, and malarial.

The majority of the people still live in the territories of their forefathers, in villages and hamlets which are basically communities of kinsmen. Some ten thousand such communities, as well as dispersed settlements, are to be found throughout the colony.

Prior to colonial contact, the culture was preliterate. Not only were there no means of recording the past, such as any form of writing, but oral tradition was limited to a few generations; and most groups had not developed a calendrical system nor more than a limited number system. Over five hundred languages are spoken in eastern New Guinea, with numbers of speakers ranging from a few score to over 60,000. Large political units such as states or kingdoms never developed: rather, the political organization was cellular, with alliances of kin groups generally embracing only one or two thousand people.

With unimportant exceptions, the people were agriculturalists, cultivating a range of tubers, other vegetables, fruits, and useful plants within their group's territory. Techniques of permanent cultivation had not been developed. There were no draught animals or beasts of burden.

In general, land rights were vested in the community: individuals had usufructuary, rather than absolute, rights in land. Cultivation rights to particular plots were allocated either by the men of status or by group consensus. There was no system of primogeniture—such rights and property as a man possessed were distributed among his heirs.

Intergroup conflict, frequently culminating in pitched battles, was perennial; but territorial conquest seems to have been fairly limited and confined to the closely settled highland regions. Groups routed from their territory usually sought to reclaim their land in further fighting, thus the boundaries of tribal and clan land were in a state of constant flux.

Most communities were able to provide from their own lands all the requisites of life. There was no cash economy, and virtually no division or specialization of labour. There was, however, a well-developed exchange system in goods and services between kinsmen and allied groups, as the principle of reciprocity pervaded all relationships between individuals and between groups. Some items, particularly certain shells, had what amounted to currency value.

These traditions and conditions therefore depict an extremely under-developed region. Moreover, Fisk has noted that 'the economic activity of the majority of the people is still conducted almost entirely outside the exchange economy. Their produce, labour, land and possessions are in the main not exchanged for money at all and have no meaningful price' (Fisk 1962: 26). Additional indicators would suggest that New Guinea might be characterized even as 'primitive' or 'pre-developed'.

## THE BACKGROUND TO COLONIZATION

When eastern New Guinea was colonized in the 1880s the colonial era was largely over in Latin America, well-established in the Asian tropics, and still being implemented in Africa. By the time Australia became a colonial 'power', therefore, imperialism in greater or lesser degree had a history of over three centuries.

Europeans were first attracted to New Guinea by the prospect of trade in tropical commodities. The Dutch and Germans, from their respective bases in Indonesia and the South Pacific, made attempts to establish settlements and trading posts in New Guinea waters early in the nineteenth century. Disease, hostile tribes, isolation, and a dearth of profitable commodities led to the failure of most of these ventures, although Europeans, and also Australians, profited during this century from trade in one commodity —Papuan slave labour—which had originally been the preserve of Moslems in the East Indies.

The permanent colonization of eastern New Guinea[1] began on two fronts towards the end of the nineteenth century. By then, German commercial houses were again interested in the region, but were unable to persuade Bismarck that the German government should add New Guinea to its Pacific territories. However, a chartered company was formed and granted full responsibility for colonial management in New Guinea. As a result of the German interest, Australia, although still a cluster of colonies, became concerned for its territorial security and urged Britain to forestall Germany by herself annexing the eastern sector. Britain's reluctance to add further to her Pacific responsibilities matched that of Germany, but her hand was eventually forced by agitation from

[1] New Guinea west of longitude 141°E. was annexed by Holland in 1848.

the Australian colonies, which included an attempt at annexation by the Queensland government in 1883. In the following year Germany annexed the better-endowed north-east region, and Britain established a Protectorate over the remaining south-east quadrant a few days later.

British New Guinea was administered jointly by Britain and the three colonies on Australia's eastern seaboard until the federation of the Australian colonies in 1901. Australia was then granted control of Papua, as it became called, but did not assume full responsibility for it until 1906. Following the outbreak of the First World War Australia occupied the German possession and in 1920 received a Mandate for its administration from the League of Nations.

Thus Australia's interest in New Guinea arose from strategic considerations rather than an interest in economic exploitation, although she had used economic arguments as well in her efforts to secure annexation by Britain in the 1870s and 1880s; the prospect of economic exploitation also arose after 1906 when Australia began to administer Papua fully. In sum, Australia gained control over her New Guinea possessions by accident rather than design: continued British control would have satisfied Australia's security requirements, and it is unlikely that she would have intervened in the German colony but for the outbreak of war.

Having secured New Guinea as a defence bastion, Australia's principal interest in the territory was fulfilled. A Royal Commission was appointed by the Australian government in 1906 to formulate a policy for Papua, and recommended a vigorous programme of white settlement and development. This proposal reinforced British precedent, namely, that colonies should be self-supporting. Policy formation thereafter was left largely to the administrators appointed to the colony, and policy implementation was perforce contingent on such revenues as could be raised internally. The small annual contribution[2] made by the Australian government before the Second World War has been termed 'official pin-money' by one writer (Stanner 1953: 15). This meant, then, that the rate at which health and education programmes, the establishment of district administration, transport networks, and the like, could be undertaken was determined by the rate of expatriate investment and commercial enterprise, the only other source of revenue.

[2] Around $50,000 before World War I; $85,000 by 1940.

This situation characterized the colony until the 1940s when Australia was threatened by Japan, and New Guinea was involved in the Pacific War. Although the colony had been acquired against just such a contingency, it was totally defenceless and unprepared when invaded.

## THE IMPLICATIONS OF COLONIAL POLICY FOR DEVELOPMENT

By 1946, the rationale for Australia's involvement in New Guinea seemed to become the reverse of the original situation. Under the extensive colonial disengagement of the period, and under the scrutiny of United Nations Trusteeship missions, the development needs increased, while the strategic value of the possession in a nuclear age declined. At least, it seemed to have waned for a period. But the strategic value was again enhanced after the Korean War was fought by conventional means, and reinforced by the flaring up of Indonesian claims to West Irian in the late 1950s. After the withdrawal of the Dutch in 1962, defence and strategic considerations again became a major part of Australia's concern in New Guinea. In a sense, however, the shortcomings of Australia's planning in this period are even more apparent than they were in the pre-war years. The damage and destruction of the war, coupled with a post-war decision by the Australian government to undertake financial responsibility for the colony's development at last, provided the opportunity to initiate a totally new programme for the colony. By this time also, the needs and problems of development were much clearer.

### *Economic Conditions*

Zimmerman concludes that economic analysis of underdevelopment is best served by reference to the per capita income of countries (Zimmerman 1965: 3-4). Fisk's observation cited earlier, that the economic activity of the majority of the people is conducted outside the money economy, suggests that the criterion of per capita income can have no wide relevance in New Guinea. However, some estimates are available for 1960[3] which provide data for comparative purposes. In that year, per capita income in the United States was almost $2,000, and half that amount in Australia and New Zealand. In 1960, the per capita income for the total population in New Guinea was about $42. This figure,

[3] Prepared by R. C. White of the Reserve Bank of Australia, and cited in Fisk (1962: 28-35).

however, takes account of both non-indigenous and indigenous income. Indigenous cash income was calculated at about 30 per cent of the gross domestic product, or in per capita terms, an income of about $12 annually. The uneven spread of this income within the indigenous sector is vividly demonstrated by the fact that the Tolai people, who number about 2 per cent of the population, had an annual per capita income of about $60; the people of the Eastern Highlands District, comprising 17 per cent of the population, had a per capita income of $0·24.

The reason for these very low per capita incomes has been suggested already: the economy is still very strongly subsistence-oriented. New Guinea is typical of most colonized territories in that only a few regions have been developed, while in the rest of the country the conditions of life and the economy are little different from those of past centuries. Only a few centres are needed to serve traditional colonial purposes—the headquarters of government, the plantation or mining districts, and their associated transportation and shipping facilities, the few cultural institutions serving mainly the alien community. Although the Administration has always been concerned with exploration, the control of tribal warfare, and the establishment of the *Pax Britannica*, the thin scatter of outposts and district stations does not negate the condition of uneven development.

Until the 1940s, a 'cores and peripheries' situation was totally applicable to New Guinea. Rowley points out that no District Office was established away from a seaport until after the Second World War (Rowley 1965: 19), and Stanner speaks of the 'fixed frontier' of pre-war activity (Stanner 1953: 15), a frontier which almost everywhere bordered the coasts. In the 1950s, however, under the 'new deal' for the colony, the central theme of policy was 'uniform development', a policy which in theory should have eroded or diminished the discrepancies between the few regions of economic activity and their extensive peripheries. However, it was rather late in the day for such a policy, when some groups had been in continuous culture contact with Europeans since the 1880s or earlier, while others in the recently-discovered highlands had either been visited only by an occasional patrol, or still remained isolated from the Administration. The policy of uniform development was further annulled by being coupled with another and older conviction—that of gradualism[4]—which persisted until it was

[4] For example, this statement by the Minister for Territories in 1958: 'The slower the growth the sounder it will be' (Hasluck 1958: 84). On the

emphatically denounced by the United Nations Visiting Mission of 1962. The end result of 'uniform development' has been to perpetuate the unevenness of development.

*The Primary Sector: the Expatriate Economy*

Heilbroner has described the economic structure of underdeveloped countries as 'lopsided' and 'distorted' (Heilbroner 1963: 75). Invariably, this condition is attributable to emphasis and reliance on primary sector production. This, too, is associated with imperialism, as the colonies were regarded above all as fields for exploitation and suppliers of raw materials.

In New Guinea the most developed regions are those with fertile and level land suitable for plantation agriculture, on or near the coasts. The exceptions are Port Moresby, which has no extensive productive hinterlands but is the administrative headquarters for the Territory; the Wau-Bulolo district in the Owen Stanley Ranges, originally opened up for gold but now the main timber-producing region; and the highland valleys where environmental conditions are so favourable for commercial agriculture that they outweigh the lack of ready coastal access. Agricultural production at present contributes nearly 90 per cent of total exports. Its dominant position in the colony's economic structure is due to the fact that the indigenous people were traditionally and almost exclusively agriculturalists, and because aliens found little else worth developing, apart from short-lived mining ventures. Under the pre-war burden of self-development and self-sufficiency imposed on the colony, its administrators were obliged to encourage and assist private investors in any enterprise which seemed promising in order to obtain operational revenues for the colonial coffers.

Although after 1945 the situation changed theoretically, with native development stressed as the central theme of policy, and financial responsibility underwritten by the Commonwealth government, in fact 'the plan still envisaged the parallel development of resources by European as well as native enterprise' (Legge 1956: 195). For example, the highlands, which contain two-fifths of the total population and which were first explored in the 1930s, were made available in the 1950s for European plantation development,

---

other hand Margaret Mead had concluded two years earlier, after field work in a New Guinea society, 'partial change . . . can be seen not as a bridge between old and new . . . but rather as the condition within which discordant and discrepant institutions and practices develop and proliferate' (Mead 1956: 204).

although forward planning might have suggested the wisdom of preserving surplus lands here for eventual indigenous development. Something over 40,000 acres has been alienated—an apparently minor amount—but some highland groups are already facing the pressure of population on their land, and many have no lands suitable for commercial agriculture. Rowley has commented on this situation:

> The climate of opinion in the Administration was still largely unaffected by what was happening elsewhere in the colonial world, and adhered to the basic assumption that the European settler employing native labour is indispensable for economic development.
>
> Probably the whole episode of white settlement in the Highlands, as a piece of administration, may be summed up as an anachronistic attempt to get capital invested in the middle of the twentieth century, in accordance with principles more applicable to its first two or three decades.
>
> (Rowley 1965: 119, 121).

At the end of the 1950s, agricultural land was still being made available for expatriate development, under the auspices of a soldier settlement scheme. When the scheme closed in November 1962, 141 European planters had been allocated blocks of land, plus loans amounting to over $6,000,000 (Howlett 1965: 9).

Alien-owned estates contribute the bulk of primary production,[5] which encompasses a narrow range of tropical crops. Some, such as coffee, are in perennial over-supply on world markets; others, such as rubber, face competition from synthetics; others again, such as the by-products of the coconut, face competition from a number of alternative products. Only negligible amounts of coffee are consumed by the local market—the rest of the production must be exported. Moreover, the value of these commodities relative to the value of manufactured goods progressively declines. The market for such commodities is characterized by booms and slumps, usually on a short cycle. But although commodity booms in the short run may compensate for periods when prices are low, the trend in recent years has been for the price of manufactured goods to rise only; thus the long-term trend is adverse for the producers of export commodities. The range of export commodities has been broadened in the post-war years by the addition of coffee and cocoa to the pre-war staples, rubber and copra, but such crops have not resulted in a more stable export economy.

[5] About three-quarters of the copra, two-thirds of the cocoa, half the coffee, and almost all the rubber.

This inadequate and unstable economic base was permitted to become re-established by expatriates after 1946, and has been actively promoted among the indigenous communities as the principal, and almost the only, form of economic development.

Yet in the fiscal year 1965-66 some $23,000,000, or over one-third the amount provided by the Commonwealth's grant to New Guinea in that year ($64,000,000), was spent on imported foodstuffs. The Annual Reports in which these statistics are contained state baldly that 'the commercial life of the Territory is based mainly on the production and sale of primary products and the importation of manufactured goods, including foodstuffs (Annual Report, New Guinea 1965-66: 61). Much of the imported food is consumed by indigenous people, both as part of the mandatory food ration for most wage earners, and through private purchase. Yet, given the colony's latitude and altitudinal range, few of the food imports could not be produced locally. Of the staple items, only cool temperate cereals must be excluded. A high proportion of the imports consists of protein items such as meat and fish, although a recent survey by the International Bank for Reconstruction and Development estimated that New Guinea has some ten million acres suitable for pastoralism (I.B.R.D. 1965: 32), and the Territory's waters are known to abound with edible fish. The production, processing, distribution and sale of these items would inculcate new skills, provide new fields for indigenous investment and employment, additional means of both saving and generating local capital, and further, ensure a more stable market for primary production than the present export-oriented complex.

Expatriate plantation development has been accommodated at the expense of over-alienation of the limited productive land, and at the expense also of the progressive disaffection of the plantation labour force.

The question of land alienation deserves some comment, especially as the Administration's land policies have been consistently labelled conservative and enlightened. By mid-1965, 3 per cent of the Territory's total area had been alienated from customary tenure, of which 1 per cent, or roughly one million acres, was classified as rural holdings. Three-quarters of the rural holdings were in the Trust Territory, the remainder in Papua—ratios which reflect both the more aggressive early policy of land acquisition in the former Territory, and also its greater agricultural potential. The percentage of alienated land appears insignificant, but in fact represents a major part of the land suitable for commercial agriculture. Although the International Bank for Reconstruction and Development estimated that about six million acres have a good potential for crop production (I.B.R.D. 1965: 67), much of this, while suitable for subsistence crops, cannot meet the requirements for commercial agriculture, which include reasonable fertility, moderate slope and altitude, assured rainfall, and accessibility to ports. It is reported that by 1940 'much of the best land had passed out of native possession . . . large areas were foreclosed to occupation and development by local natives who were short of land' (Stanner 1953: 33).

Not only is the dispossession of land serious, but also the fact that much alienated land lies undeveloped. From the 1890s in Papua, and after 1921 in the Trust Territory, land legislation was designed to permit alienation only through the agency of the Administration, which provides 99-year leases to successful applicants. The lease contracts contain clauses providing for the resumption of land which is not developed within a specified term; but these provisos have been rarely invoked. Thus, of 93,000 acres leased to mission societies in the Trust Territory by 1940, only 29,000 acres were under cultivation (Stanner 1953: 33); of 19,000 acres in the Highlands leased for coffee plantations by 1961, the actual planted area was around 6,000 acres (Rowley 1965: 120).

I have commented elsewhere on the labour situation on the plantations (Howlett 1967: 83ff., 141ff). The critical issues to note here are two. First, the main plantation districts are not synonymous with the regions of greatest population, hence migrant labour has been inevitable on the plantations. The conditions of wages and accommodation, however, are such that only males, married or not, can move to the plantation districts for employment. Highlanders form the bulk of the plantation workforce, and must be repatriated after two or three years' employment. The system is not only uneconomic for the plantations, but more seriously, disadvantageous to the villages: they are deprived of able-bodied men without adequate compensation; family members become indebted to kinsmen and friends for services which the absent workers would normally undertake; and families are prevented from making a stable and permanent transfer out of the subsistence milieu. Second, employment as a plantation labourer does not automatically improve a person's ability to produce the same crops in his own village plot, and the supposed 'demonstration effect' of plantation labour is nil for highlanders, who cannot grow lowland crops in their villages in any case.

The adverse economic and social effects of the plantation system must be imputed to the government rather than the planters. Initially, as we have seen, Australia's totally inadequate grant obliged the Administration to rely on locally-generated capital, and plantations were an obvious source of such revenue. Legislation has been devised to assure planters of land and labour at the ultimate expense of the villagers. The opportunity for at least diminishing the reliance on plantation production was provided by the war, but not taken. Finally, the same economic programme, that is, the production of export crops, was transferred to the indigenous sector. This programme, adopted in the name of 'economic development', amounts to little more than following the lines of least economic resistance.

*The Primary Sector: the Indigenous Economy*

Papua's first administrator, Sir William MacGregor, envisaged the growth of an independent village economy, with villagers participating in the cash economy by the production of the typical export crops. His policy was intermittently endorsed by subsequent administrators in both Territories until the 1940s,[6] and vigorously promoted in the post-war years. The limitations of such a programme for a territory facing the prospect of independence scarcely seem to have been recognized even now by the policy-makers. The acute issues affecting the subsistence sector, which concern the agricultural technology, the types of crops comprising the staples, and the traditional land tenure systems, have scarcely been dealt with. Instead, cash cropping has merely been grafted on to the existing subsistence base, with all its inadequacies. To modernize agriculture in New Guinea, the agricultural foundations must be rebuilt; modernization, in effect, demands an agricultural revolution, transforming the technology of production, the staple food complex, and customary land tenure. Drastic measures indeed, but indispensable for eventual development and growth not only in the primary sector but of the entire economy. The need to transform and modernize the primary sector is seen by every economist of under-development. The economic condition which demands such a transformation, and which typifies New Guinea, is usually summed up thus: '. . . a low level of productivity is the

[6] In 1918 a system of compulsory cultivation was instituted with the passage of the Native Plantations Ordinance to contribute revenue for the Administration. The system lapsed during the Second World War and was not renewed.

universal economic attribute of underdevelopment, a fearful shortage of capital its well-nigh invariable cause' (Heilbroner 1963: 56). As over 90 per cent of the population is wholly or partly engaged in subsistence production, the overall economic situation is far from satisfactory. A diversified and better-balanced economy demands that a substantial part of the workforce move out of the primary sector into the more productive, capital-generating fields, mainly in the secondary sector, and that those remaining in the primary sector produce both more and different foods. These issues will be treated in turn.

Nowhere in New Guinea are subsistence fields under permanent cultivation. It is not even certain, if rigorous experiments by Belgian agricultural scientists in the Congo are indicative, that low-altitude terrain in these latitudes *can* be brought under permanent cultivation without controlled irrigation and massive fertilizer inputs (Pelzer 1958: 129). While shifting cultivation represents the essential technology of food production, large areas of land are necessary to the maintenance of soil fertility.

The apparent abundance of land in the Territory conceals some facts about the availability of cultivable land. As noted, neither the population nor the land resources are evenly distributed; moreover, much land is kept out of cultivation because the rights to it, after generations of intertribal conflict, are uncertain or unknown; land is also kept out of cultivation by taboos and the fear of sorcery. At present the population is small enough to preserve the critical proportions between cultivation and fallowing, but shifting cultivation can support neither a large population nor a population experiencing rapid increase. So far, the only important technological change which has taken place in traditional agriculture has been the substitution of steel tools, especially the axe and spade, for the stone axe and digging stick, a substitution which is now very nearly universal in the villages. The 'saving' in man-hours resulting from the use of steel tools, combined with the prohibition on warfare, has made production more efficient, but has not led to significantly greater food production for the reason that there is no outlet for it. Some of this 'saving' has been invested in cash crop production, but many village men now have more time on their hands than traditionally, and fewer means of employing it. The staple foods are partly responsible for this situation.

In New Guinea, as in much of Africa and equatorial America, the staple foods are root crops and tubers, frequently comprising 90 per cent of the diet. New Guineans can produce a surplus, but they cannot store it and accumulate it, as tubers deteriorate fairly rapidly. Traditionally such surpluses were disbursed in periodic feasts between friendly groups, to cement alliances, to celebrate weddings, and the like. Surpluses are needed now less for these purposes than to supply the needs of those in paid employment, and to reduce the colony's heavy expenditure on food imports.[7] Production of a storable surplus would not only supply locally-produced foods to non-agricultural workers, but also help mitigate seasonal crop failure and regional calamity. Moreover, tubers have low nutritional value, and need to be replaced for the improvement of the health and energy of the population. A considerable social as well as economic problem is posed by the introduction of new staples, but the challenge must be met.

Changes in the communal land tenure system would appear necessary for the rationalization of both food production and commercial agriculture, principally because the security provided by private land ownership gives an important incentive to entrepreneurship. Traditional land tenure systems pose more problems for the cultivation of tree crops, which have an economic life of upwards of a generation, rather than a season or two. Unless land reform is accomplished which will transfer to the individual security of tenure to a block of reasonable size, and prevent its subsequent fragmentation, there is scarcely any point in cash cropping for the villager.[8] Without these protections cash crops are grown in the same manner as the subsistence foods, in small, scattered plots of land, susceptible to claim by other kinsmen, and yielding a pittance.

Customary land tenure also inhibits more rational land use, thereby depressing output, for food production. For example, in the highlands where soil fertility can be maintained under cultivation for longer periods than in the lowlands, greater productivity could be achieved if the cultivable lands of the villages could be merged into areas large enough to be mechanized, especially for the arduous tasks of clearing. The present system, with families or kinsmen cultivating small scattered parcels of land, is patently wasteful of time and effort, although it was compatible with traditional socio-economic organization.

[7] The export of some subsistence crops, such as bananas, pineapples, and sugar cane, is precluded by Australia's production of these items.

[8] For an analysis of the prospect of New Guinea becoming a nation of peasant proprietors with a small-holder economy, as envisaged by post-war policy, see Howlett (1967: 124).

The addition of long-term tree crops to the short-term subsistence crop complex has meant that most of the cash crops compete with subsistence foods for land; cash crops therefore produce a similar result to an increase in population—they reduce the amount of land available for food production. Thus the present adequacy of land may soon pass, as both the population and the area planted to tree crops are increasing. If the problems outlined here are not in the process at least of solution before all the arable land is committed, then the outlook for New Guinea will be bleak indeed. When Mr Hasluck was Minister for Territories he observed:

> One of the worst things that can happen to the individual is to educate him to live at a higher standard and to require a greater number of possessions and then to deny him the means to do so.
>
> (Hasluck 1958: 106).

But this is precisely what seems to be happening, and under present trends little prospect of alleviation is in sight. The frustrations resulting from this situation have found outlet among the Territory's people on numerous occasions during this century in the unjustly notorious 'cargo cults', messianistic reactions to stress which have also been manifested in other parts of the colonial world.

Legislation enacted in 1962 and 1963 provides for the conversion of land throughout the Territory to individualized tenure; but rationalization of customary land tenure proceeds slowly. The problems are recognized to be complex but not urgent, thus there are less than a score of Land Commissioners (one for each of the eighteen Districts) to deal with the multiplicity of claims, disputes, and investigations involved.

One further observation must be made about the export crops grown by the villagers. In addition to the problems of adapting such crops into the traditional socio-economic system, or, more to the point, of adapting the system to the new economic requirements, there is the complication that these crops are produced for foreign markets. The difficulties encountered by European producers, such as unstable prices and uncertain markets, are compounded for the village producer whose production is usually lower in yield and in quality, and who is generally unaware of the international scale of production or the international mechanisms for marketing these commodities. This is not to argue for the abandonment of export production: the objection is not so much to the commodities involved, as to the virtual total. reliance on such production for development. Gourou has argued that, for developing countries,

> The essential thing . . . is that the export of agricultural products should take place; better to have commercial crops produced by poor techniques than an agricultural system that is not oriented towards exportation. . . . This export cropping resulting from local initiative offers the possibility of gain, gives rise to a trading class, and presents opportunities for mastering improved techniques.

He adds an important proviso, however:

> A necessary condition for the development of this kind of native initiative is the absence of large estates and of monopolistic foreign enterprise which deprives the natives of the opportunity for technical advancement that cultivation for export offers.
>
> (Gourou 1966: 176).

*The Secondary Sector*

The secondary sector in New Guinea is only embryonically developed. Most 'industry' involves the preliminary processing of agricultural commodities for export, and engages a relatively minor segment of the workforce. Fisk has estimated (Fisk 1966: 28) that of the indigenous people who earn a cash income, only between 10 and 14 per cent do so in industries other than agriculture. In 1965, of a total paid indigenous workforce of over 91,000, only 7,835 were employed in factories[9] and only 648 were receiving training as apprentices (Production Bulletin 1965). About 70 per cent, or some 64,000 of those in paid employment, had unskilled labouring jobs on the plantations or in the towns.

These statistics emphasize the need for skilled and semi-skilled employment, and the necessity to develop the secondary sector. The majority of the paid workforce has only the most precarious toehold on the cash economy. Most of the people in this bracket would be forced back on subsistence in the event of economic recession, or reduced Australian investment, as may well occur after independence. The Papuanization of virtually all spheres of economic activity has been so slow that the International Bank for Reconstruction and Development, in drawing up a development plan for the country, acknowledged that: 'if the Territory were left

[9] In New Guinea, defined as an industrial establishment employing four or more persons, or using motive power other than hand power.

to its own means there would not be the technical skills, the management or the finance to develop the economy at any reasonable pace' (I.B.R.D. 1965: 31).

An adequate infrastructure has yet to be established. Fisk has predicted (Fisk 1962: 39) that the cost of providing the necessary infrastructure for economic progress will be 'frightening', but added, 'in the absence of an unexpected bonanza, such as the discovery of rich oil fields, *there is no other path to a viable economy*'.

Given the conditions of terrain, the lack of conventional industrial resources, and the small and scattered population, overwhelmingly illiterate and unskilled, the feasibility of developing the secondary sector would seem to be limited. These conditions certainly proscribe the development of heavy industry complexes for the foreseeable future. On the other hand, a programme aimed at a modest level of industrial activity, especially in the field of import replacement, is not only feasible but essential. The processing and marketing of the production from an enhanced and diversified primary sector has been suggested. Other items which would round out both the primary and secondary sectors are the production and processing of industrial fibres, for example, jute, abaca, kenaf or sisal, for a variety of household items and particularly to supply sacking for the exported commodities, and possibly cotton for a textile industry. Manufacture of footwear and other items will be possible once a pastoral industry is established. And so on. The possibilities are multiple.

The potential is also greater than at first appears. An essential precondition for economic 'take-off' is the accumulation of capital for investment in developmental projects. That this stage has been approached in New Guinea may seem unlikely, but even at the present low level of economic development capital is being saved and accumulated.[10] In June 1966, for example, nearly 214,000 indigenous people held savings bank accounts totalling 9·2 million dollars; considerably greater amounts are known to be hoarded in the villages, a fact made apparent in 1966 with the introduction of decimal currency, requiring conversion of the former currency. But many bank accounts are not operated: local banks periodically advertise lists of these 'dead' accounts. Such stagnant capital is useless; to assist economic progress it must be activated and used

[10] 'Saving' in the sense of freeing labour from consumption-goods production has already occurred, but very little has yet been applied to capital-goods production.

productively. That the resources of indigenous labour and capital are not so used implies a lack of incentives and of fields for investment.

*Social Conditions*

It has been stated that 'The first requirements for high labour productivity under modern conditions are that the masses of the population shall be literate, healthy, and sufficiently well fed to be strong and energetic. In many countries . . . if this were achieved all else necessary for rapid economic development would come readily and easily of itself' (Viner 1953: 82).

*Education*

As noted, New Guinea was preliterate when colonized. To an overwhelming extent it still is. Until a 'crash programme' of education was initiated in 1960, popular education may be said to have claimed the least part of the government's attention. Stanner is of the opinion that native education 'from the beginning, had been looked upon as a privilege of the missions rather than a duty of the government' (Stanner 1953: 30). This statement is somewhat exaggerated, for some at least of Papua's administrators were anxious to provide more education facilities than their limited budgets would allow, but it broadly represents the position of the Australian government until recently.

The Annual Reports for the Territories contain no data on literacy rates, an omission which doubtless indicates that literacy levels are still very low, especially in English. Literacy in the *lingua franca*, Pidgin English, has often been acquired informally and is more widespread, particularly among men. Not all members of the House of Assembly understand English; the business of the House is conducted in English, with translation services.

The neglect of education has been remedied belatedly. Since 1960 primary schools of several standards have been widely established. The great majority follow a syllabus which is 'specially designed for Territory pupils', and emphasizes 'the best elements of indigenous culture', particularly music, art, handicrafts, dancing, social studies, sports, agricultural principles, woodwork and craftwork. The remainder use the primary school syllabus of New South Wales (Annual Reports). These are now being supplemented by secondary schools, teacher training colleges, and a university. Agricultural training institutes have been increased in the same period.

Although the number of formal institutes of education has been increased in recent years, gaps persist. Training in industrial skills is minimal; much of what is available is sponsored by private enterprise rather than the Administration. There is still little opportunity to develop managerial skills. The paternalism of the Administration, however well-intentioned, is largely at fault for this deficiency. There has been a marked reluctance to admit indigenous people to managerial and decision-making spheres. The same condition is manifest in the preference for employing expatriates in many situations, rather than training indigenous people or indeed, employing those who already have appropriate training. This is dysfunctional in a second sense, as expatriates expect and are given considerably higher wages and better living conditions, an expenditure which might be diverted into other channels.

*Health and Nutrition*

While much remains to be accomplished, the concern for and extension of medical services in New Guinea has been considerable. Stringent quarantine measures have protected the colony from many tropical scourges such as smallpox, typhoid and cholera. Malaria, however, is endemic except at high altitudes. Its control poses vast problems, given the scattered population and rugged, densely-vegetated terrain. Where this disease is prevalent, it represents a serious economic liability, debilitating those affected and causing high absenteeism, as well as high mortality rates. Gunther has noted: 'Control malaria and you will double the expectation of life, halve the infant mortality rates, and double the population in seventeen years' (Gunther 1958: 48). Gastro-enteritis and dysentery are among the principal causes of death, reflecting poor standards of hygiene. Respiratory diseases affect large numbers, especially in the highlands.

Nutritional problems were briefly referred to in the discussion of subsistence. Protein and vitamin deficiency is widespread. In the villages, this arises from the predominantly vegetarian diet; elsewhere the high cost of imported foods precludes an adequate diet for the Papuan. The biggest single contribution to improving the dietary status will come from the development of a pastoral industry and the fish resources.

*Demographic Conditions*

As far as it is possible to establish from the recency of records and incomplete statistical data available, the population is increasing at the rate of about 2 per cent per annum. It is believed that some groups, particularly those with the longest history of European contact, and thus reasonable medical facilities, are increasing at a faster rate. Long-term trends are difficult to estimate, but the population may be expected to continue its upward trend, now that warfare has been prohibited, and as medical services are extended.

This situation would be encouraging if the colony were to achieve diversification and expansion of the economic base: a small population (especially when characterized by low per capita income) is a liability to economic growth. However, an increasing population can only be regarded as threatening while the present dependence on agriculture continues. Further, it has been estimated (McArthur 1966: 113) that under present trends, between 40 and 45 per cent of the population will be under fifteen years of age by the mid-1970s, representing a requirement for primary education for between half and three-quarters of a million children.[11] These two illustrations alone point to the urgency of the need for more industrial and education opportunities.

*Political Institutions*

It is, of course, impossible to predict the form of political institutions which will be adopted in New Guinea when it becomes independent. Australia has attempted to pre-determine that these shall be basically of the kind established in the Western democracies. But the ability of such institutions to function in a developing country characterized by economic and social disparities has rarely been demonstrated. Typically, the political institutions of underdeveloped nations are unstable and ineffective.

Many of the recently-independent, developing nations are the creations of expediency, chance, or the balance of power in Europe at a particular period, rather than ethnically and culturally homogeneous units. It has been shown that the present political boundaries in the New Guinea region were set by such circumstances. A sense of nation, or a common recognition of the need for mutual effort to achieve developmental goals, is seen to be essential as much in the advanced countries as in the developing ones. However, fostering national identity in the once-dependent territories, with their lack of conformance between political boundaries and cultural affiliation, has proved difficult except by radical and extreme methods.

[11] Some 250,000 children were attending primary schools in 1965.

The decades of Australian administration so far have made few inroads in changing the traditional political allegiances. Rowley observes that 'the concept of "New Guinea" is still unfamiliar to many' (Rowley 1965: 15) and the United Nations Visiting Mission of 1962 claimed that New Guinea's main problem was that of 'creating a single people with a common purpose'. A sense of nationalism in the years preceding and following independence is regarded by indigenous leaders throughout the developing world as a necessary step in political maturation. Often it is a step made more complex by the 'unnatural' boundaries of the embryonic state, and by the existence within it of a plural society.[12] But in many senses nationalism is now an obsolete ideology, both politically and economically. National independence is at best conditional in respect of the contemporary structure of economic, political and military power. Thus, while the preoccupation with national identity may be inevitable, at the same time it is often an impediment to the course of development.[13] If this is so, then New Guinea is at a disadvantage and is not benefited by the lateness of its colonization.

The political experience of the population will be relevant. The establishment of a House of Assembly in 1964, elected from a common roll of adults, with an indigenous majority, was due very largely to the insistence of the 1962 Visiting Mission. Until then, the most important steps toward native participation in their own affairs had been at the level of the local government councils, first established in 1950. This is not to deride the great value of the councils, but to indicate the limits of official sanction to indigenous participation in the country's affairs. Expatriate guidance is essential in the initial stages of organization, be it for political or economic purposes. But the pre-emption of authority by the alien officers-in-charge has always been characteristic of such organizations in the colony. The highest status a Papuan may attain in the House of Assembly is that of Under-Secretary; and final authority for the passage of ordinances still rests with the Australian government.

[12] Furnivall's definition of the term is appropriate here, namely, 'A society comprising one or more elements or social orders which live side by side yet without mingling, in one political unit' (Furnivall 1944: 446).

[13] For example, the events of the early 1960s in the Malay world, which included Sukarno's drive to 'liberate' West Irian and his 'confrontation' adventure; and the schism between Singapore and Malaya. The cost of national status symbols, such as 'show-case' industrial complexes and national airlines, may also be included here.

P

The social structure of a nation profoundly influences its political character. Social pluralism, in the sense referred to above, is entrenched in New Guinea. Between the indigenes and the Europeans is a third social group, the Asians, most of whom are descendants of Chinese artisans and labourers brought in during the German regime. Australia prohibited the entry of Asians to Papua, applying to it the terms of the 'White Australia' policy, and did not permit Asian immigration to the German colony after the Mandate was obtained. Asian residents of the Trust Territory were only recently permitted residential status in Papua. Although the Asian community is small, numbering only about 3,000, eventually its economic strength in commerce and plantations may earn for it the same resentments and reactions as have been manifested against similar groups in South-east Asia. That such a reaction could occur seems to have been recognized by the Minister for Territories a decade ago, although his remedy is of questionable merit: he expressed the need to 'attract the Asian and mixed-blood elements in the population towards the European population so as to avoid the perpetuation of too many racial groups' (Hasluck 1958: 77). While this end may be achieved, it cannot but serve to perpetuate the separateness of the Europeans and indigenes.

Ethnic separateness is most obvious in the Territory's towns. The 1966 census revealed that almost 104,000 indigenous people now live in the urban areas. The census defines an urban area as any (non-traditional) community with a population of 500 or more. The non-indigenous population of the urban areas is about 25,000. In the towns the discrepancies in status and wealth between the two sectors are sharp. Aliens enjoy a higher standard of living than they would in Australia, with allowances, higher salaries for equivalent occupations, tax benefits, and domestic servants. Indigenous people on the other hand undoubtedly have a poorer standard of living than is available in the villages. By no means all of the indigenous urban dwellers are employed—many have drifted to the towns in the hope of finding employment, but are denied it by having no skills to offer. Squatter settlements have been part of Port Moresby's urban environment for some years. The indigenous immigrants are predominantly adult men, who may be tied to the town because they lack the resources for the journey back to the village, or who may no longer find custom-bound village life satisfying. Thus, the current situation represents not only economic waste but a social condition which has already become critical in the larger towns. Over the last decade or so the more blatant instances of racial discrimination, ranging from a prohibition on alcohol, to segregated clubs, beaches and cinemas, and exclusion from political institutions and the bureaucracy, have been removed. Even so, many less overt instances of pluralism persist. In this context, Barbara Ward's comparative depiction may be noted:

> . . . there are perhaps only two types of ex-colonial community in which it is excessively difficult to achieve a transfer of power with anything like the goodwill which is needed to make it effective and peaceful. One type is the country where a settler problem complicates the issue. . . . In such communities the lines of cleavage between groups—the political, social and economic lines—are strengthened and exacerbated by the greatest dividers of all—the dividers of race and culture. Settlers from the metropolitan country come in and root themselves in the local community. They take the best land. Being better educated, they produce more wealth from it. They hold the best posts. They often control the administration. . . . Two societies develop. In one, the white settlers build up a more or less wealthy modern community. Around and among them the dispossessed exist, multiply, and finally begin to revolt. . . . Here the transfer of power presents overwhelming difficulty.
>
> (Ward 1962: 120).

One final example of ethnic exclusiveness may be added. The inhabitants of the Trust Territory have the national status of 'Australian Protected Persons', but those born in Papua are legally Australian citizens; citizenship does not, however, permit the Papuans to enter at will into Australia, either for employment, residence, or even vacations.

While ethnic pluralism in broad terms does not characterize the indigenous population, economic and social pluralism are found within it. The pluralism which arises from unevenness of development and experience has been referred to already. To this may be added the acute differences which distinguish the experience of men and women in the Territory. Prevented by employment conditions from accompanying their men to other parts of the country, and often obliged to undertake their husbands' tasks in the villages, the women of the Territory remain very much in the tribal condition.

The complex of issues outlined here does not augur well for independent political status, yet such status is inevitable, and probably imminent. The postponement until 1964 of native admission to the country's governing body has restricted political experience; the exclusiveness of the alien bureaucracy, which persisted until

1964, has deprived the indigenes of requisite administrative skill; the rapid and demographically unbalanced urban immigration is hastening detribalization and discord; pluralism fosters élitism; sectionalism will undoubtedly colour all political action for some time to come. Thus if New Guinea is to adopt the political institutions of Western democracies on independence, it would seem that these institutions must inevitably be unstable and ineffective.

## CONCLUSION

What conclusions may be drawn from this survey? One must note, again, the considerations presented initially. Colonialism and under-development are well understood. Their characteristics, tendencies, and consequences have been extensively explored and documented. Australian policy in New Guinea is applied in this context of understanding.

Enough has been said to demonstrate that New Guinea derived no advantage from having been among the last of the colonized territories; nor from having become the dependency of a nation so recently released itself from colonialism. No significant aspect of Australian policy reflects the lessons of the past, or of the under-developed world. Australia's access to generations of colonial policy and practice might have yielded approaches and programmes which aimed to prevent or circumvent the more unfortunate consequences of colonialism; consequences which now compound the problems of development in former dependencies. Far from recognizing the weaknesses and deficiencies of the record, Australia has followed largely without question the precedent of other colonial powers; and in fact has duplicated in the span of a few decades the characteristic features of colonialism which evolved during more than three centuries. New Guinea, it seems, will have to bear the consequences in its modernization efforts in much the same way as have other colonies.

Explanation of this blindness does not come readily. The motives for colonization have been outlined. Possibly one should make allowance for some circumstances of Australia's own development during the past century. For example, at the end of the nineteenth century the Australian colonies were experiencing economic depression, widespread strikes and unemployment, which may well have played a part in relegating New Guinea to the limbo it was to occupy for the next half-century. Then again, there was already a strong feeling against coloured peoples—a feeling which seems to have arisen during the mid-century gold rushes, directed against

Chinese immigrants; and which mounted again later with objections to Queensland's traffic in Pacific Islanders for labour in its cane-fields. This sentiment was embodied, with federation, in the Commonwealth's immigration policy, as the so-called White Australia policy. Such factors may have conditioned Australia's early attitudes to New Guinea, and the wars and inter-war depressions of the twentieth century may have helped entrench them. Stanner suggests some other factors:

> . . . the Commonwealth's protracted security from foreign threats, an unconscious dependence on Britain's protection, the dislike of the new epithet of 'imperialism', and cultural isolation at high standards may have had much to do with the dissociation. Perhaps also the lost sense of frontier, the galloping urbanism, and the intellectual confusion of the period [1920s and 1930s] about colonies, may have made their contribution.
>
> (Stanner 1953: 1-2).

It is, of course, unfair to judge Australia's performance in the past by the standards of present knowledge and experience. But this essay has focused attention on planning in the post-war years. In these decades, Australia has been so heedless of the accumulated colonial experience that it seems almost euphemistic to characterize its management of New Guinea as 'policy' at all.

The characteristics of under-development are frequently depicted as a vicious circle, an interlocking cycle of economic, social, demographic, political, and other problems. The omissions and commissions of Australian policy have in no way breached New Guinea's vicious circle of conditions perpetuating under-development. Rather they replicate the exploitation and neglect of colonialism, earlier and elsewhere. As Swift perceived, 'None so blind as they that won't see.'

## BIBLIOGRAPHY

BICKER, W. *et al.* (1965). *Comparative Urban Development: An Annotated Bibliography*. Amer. Soc. for Public Administration, Washington D.C.

COMMONWEALTH OF AUSTRALIA (1964-65), (1965-66). Report to the General Assembly of the United Nations on the administration of the Territory of New Guinea. Canberra.

COMMONWEALTH OF AUSTRALIA (1964-65), (1965-66). Annual Report of the Territory of Papua.

COMMONWEALTH OF AUSTRALIA (1967). *Year Book* of Australia No. 53. Bureau of Census and Statistics, Canberra.

FISK, E. K. (1962). 'The Economy of Papua–New Guinea', in Bettison *et al., The Independence of Papua–New Guinea.* Angus and Robertson, Sydney.

FISK, E. K. (1966). *New Guinea on the Threshold,* A.N.U. Press, Canberra.

FURNIVALL, J. S. (1944). *Netherlands India.* Cambridge. 2nd ed.

GUNTHER, J. T. (1958). 'The People', in Wilkes (ed.), *New Guinea and Australia.* Angus and Robertson, Sydney.

GOUROU, PIERRE (1966). *The Tropical World.* Longmans, London. 4th ed.

HASLUCK, PAUL (1958). 'Present Tasks and Policies', in Wilkes (ed.) *New Guinea and Australia.* Angus and Robertson, Sydney.

HEILBRONER, R. L. (1963). *The Great Ascent.* Harper and Row, New York.

HOWLETT, D. R. (1965). 'The European Land Settlement Scheme at Popondetta', *New Guinea Research Unit Bulletin* No. 6. A.N.U., Canberra.

HOWLETT, D. R. (1967). *A Geography of Papua and New Guinea.* Nelson, Melbourne.

I.B.R.D. (1965). *The Economic Development of the Territory of Papua and New Guinea.* Johns Hopkins, Baltimore.

JONES, G. N. (1964). *Planned Organisational Change: A set of Working Documents.* Univ. of Southern California, Los Angeles (mimeo).

LEGGE, J. D. (1956). *Australia's Colonial Policy.* Angus and Robertson, Sydney.

MCARTHUR, NORMA (1966). 'The Demographic Situation', in Fisk (ed.) *New Guinea on the Threshold.* A.N.U. Press, Canberra.

MEAD, MARGARET (1956). *New Lives for Old.* William Morrow, New York.

MEZIROW, J. D. (1963). *The Literature of Community Development: A Bibliographic Guide.* Department of State, Washington D.C.

PELZER, K. J. (1958). 'Land Utilisation in the Humid Tropics: Agriculture', *Proceedings,* Ninth Pacific Science Congress, 1957, Vol. 20.

ROWLEY, C. D. (1965). *The New Guinea Villager.* Cheshire, Melbourne.

STANNER, W. E. H. (1953). *The South Seas in Transition.* Australasian Publishing Co., Sydney.

SPITZ, A. A. AND WEIDNER, E. W. (1963). *Development Administrations: An Annotated Bibliography.* East-West Center Press, Honolulu.

TERRITORY OF PAPUA AND NEW GUINEA (1966). *Production Bulletin No. 7,* Parts I and II. Bureau of Statistics, Konedobu.

VINER, JACOB (1953). *International Trade and Economic Development*. Clarendon Press, Oxford.

WARD, BARBARA (1962). *The Rich Nations and the Poor Nations*. W. W. Norton, New York.

ZIMMERMAN, L. J. (1965). *Poor Lands, Rich Lands: the Widening Gap*. Random House, New York.

D. W. MEINIG

# *A Macrogeography of Western Imperialism: Some Morphologies of Moving Frontiers of Political Control*

*'. . . it seems necessary that in an age of intense specialization, historical geographers should sometimes scan the whole picture . . .'*

A. GRENFELL PRICE *The Western Invasions of the Pacific and its Continents*

## PREFACE

THE PRESENT GENERATION has been given the rare opportunity of viewing one of the great movements in human history as an episode, as a single vast event essentially completed in our own time. For it is now witnessing the last phase in the dissolution of a set of global geopolitical patterns, the origin of which extends back at least five centuries: the outreach of Europe and the impress of its political power for some considerable period upon very nearly every part of the non-European world.

To see that movement as an episode is, of course, to take a panoramic view which merges a myriad of lesser movements into a single grand process: Europeans of whatever origin, impelled by whatever motivations, imposing themselves by whatever means upon the political lives of other peoples. It is a view of 'Western imperialism' as a kind of historical entity, a great field of study in itself, and we may confidently expect a growing literature which will treat it as just that. An excellent example, both in its content and in the fact that it is one volume in an integrated series on world history, is D. K. Fieldhouse, *The Colonial Empires* (Weidenfeld and Nicolson, London, 1966), which the author offers as 'a prospectus for a virtually new field—the comparative study of empires' (p. xiv).

The present essay is a much smaller offering in the same spirit. It is certainly no more than a prospectus, a kind of simple framework to illuminate certain dimensions of that vast history. The essentials of that framework are displayed in the full title, but as usual in the scholarly world, the terms themselves are by no means self-explanatory. I use 'macrogeography' to specify a broad, essentially global, view of areal patterns. By 'Western' I refer to the general culture and the group of nations which developed out of European Christendom. By 'imperialism' I mean the reaching out by one people to impose some degree of political control upon another. It is now very common for 'colonialism' to be used in the same sense (as, indeed, by Fieldhouse), but I prefer 'imperialism' as a more purely political term which, to use Grenfell Price's euphonic distinction, embraces both 'sojourner' and 'settler' colonizations. 'Western imperialism' thus, for example, will include all of the eastward expansions of Russia into Siberia, but not its westward expansions into Poland for the latter is, by this definition, an intramural imperialism. So, too, I have arbitrarily excluded the imperial expansions of Japan even though they were co-ordinate in type and time with later stages of Western imperialism.

The geographical study of such historical movements can take various forms. We may display geography upon the scaffold of history, as it were, by focusing upon areal changes within a rigid temporal sequence. Or we may give primary emphasis to the areal patterns themselves, studying them as frameworks within which the processes of history take place. As the subtitle suggests, I have attempted some combination of these, but with the chief emphasis upon the latter, upon the areal forms or 'geographical morphologies' apparent within a general sequence of imperial expansions. Such expansions were the chief political dimension of those 'moving frontiers' which have been such a grand theme of so much of Grenfell Price's work. The vast sweep of the essay will at least match in audacity, if not in authority and eloquence, his own illuminations of our world, and I dare to hope that he will accept it as a complement to his own efforts.

## PRELUDE IN EUROPE

In the broadest view, the Western Imperial Era is usually considered to have begun about A.D. 1500, to be generally synonymous with that Expansion of Europe which began with the Voyages of Discovery. As historians know much better than anyone else, such periodization is never clear-cut and self-evident (some would, for example, point to the Crusades as the opening phase), but for purposes here it is less important to tie the movement to a particular date or initial event than to relate it to some patterns of expansion which were evident within and along the immediate margins of European Christendom just prior to the outreach to other continents (Figure 25).

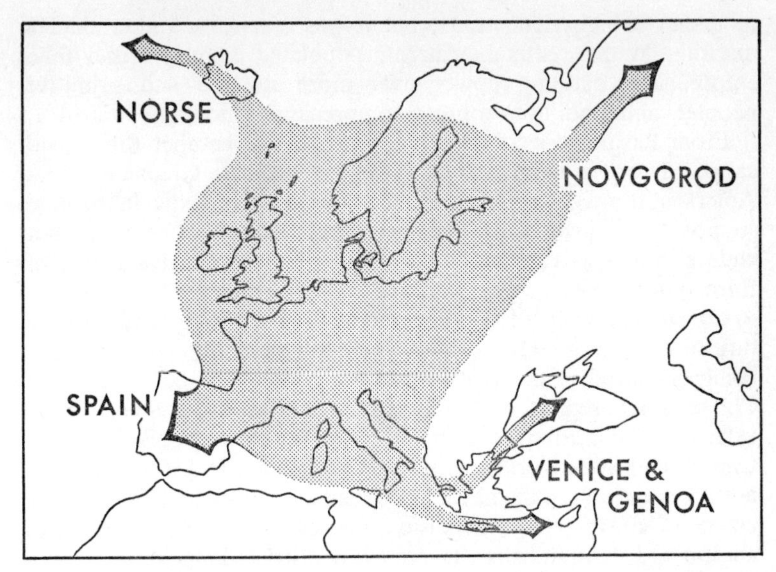


FIG. 25 Thrusts from European Christendom on the eve of the fifteenth century

Among the most striking of such patterns were the politico-commercial networks of Genoa and of Venice, reaching a thousand miles and more across the seas to outposts in the Levant, the Aegean, and the Black Sea, connecting with all the great trade routes of Western Asia. Each was a system designed to span the seas for commercial gain with the greatest efficiency, a network of islands, peninsulas, and ports bound together for protection, provisions, and trade. Together these sea empires, created by purchase and by power, commercial and aggressive, competitive with one another for the same regions and products, represented a vigorous thrust out from the body of Europe into alien ground.

Far to the north-east another thrust carried other Europeans well beyond the bounds of their homeland. For Novgorod, the main eastern inland depot of the Baltic trade, had created its own special hinterland to the north and east in the lands of the Finns, spanning the boreal forests to the Arctic seas, reaching out successively from one river system to the next, to the Dvina, the Pechora, and ultimately across the northern Urals into Yugra on the Ob. It was a thrust and system not unlike those in the seas to the south-east of Europe, a commercial network of posts and routes to gather a high-value trade, but it was a riverine rather than a maritime system, extending across populated country rather than empty seas, exerting power over more isolated and primitive peoples, and leading to a more comprehensive political control.

From the north-west corner of Europe still another thrust had carried Norse traders and colonists to Iceland, Greenland, and America. It was a network flung across seas so difficult, to land so poor, and peoples so primitive that the full system did not endure, but it was certainly expressive of the expansive power of European peoples.

Concurrent with all of these far-ranging maritime and riverine thrusts a very different exhibit of that power of expansion was being displayed in the south-western corner of the continent. The Christian reconquest of Iberia was a contiguous, comprehensive extension of control over lands and peoples, a drive impelled more by Church and State than the profits of commerce, and it forced upon non-Christians (usually never more than a minority in these lands) a choice of conversion or expulsion, and carried with it an influx of additional Christian colonists into every district. Thus, rather than a lengthy outreach across or into alien lands, it was more a slow relentless geographical growth of the main body of European culture.

Furthermore, rather similar contiguous expansions by conquest and colonization were apparent along other frontiers: on the west by the English against the Welsh and Irish; on the north by the Swedes against the Lapps and Finns; on the east by the Slavs gingerly along the edge of the Steppelands. While some of these were technically intramural in Christendom, all were a further display not only of the power but of types and techniques of expansion which would soon be applied more obviously as imperial systems. Thus we can see in the fourteenth and fifteenth centuries a centrifugal pressure outward from the core of European Christendom of which the famous Expansion of Europe of the sixteenth and seventeenth centuries is, in general, simply a later phase in larger scale, a shift from marginal seas and backcountry marchlands to the world ocean and continental interiors, a new global manifestation of earlier more local movements, an extension not an invention.

## THE CONTINENTAL EMPIRE OF SPAIN

There is no more striking exhibit of such a continuity and expansion than the sudden outburst of Spain into a wider world to create a vast imperial system.

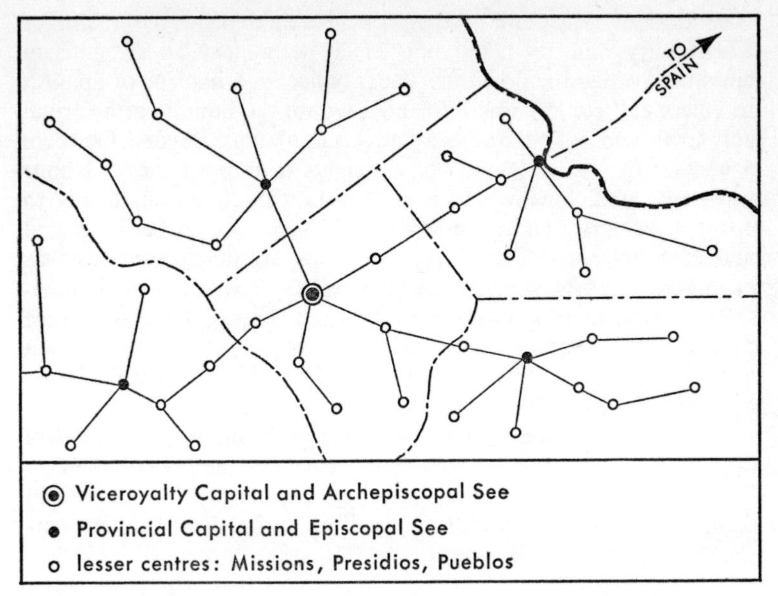


FIG. 26 Hypothetical patterns of the continental empire of Spain

Within half a century of the voyages of Columbus, not only had the Spanish reconnoitred much of the coastlands and heartlands of both Americas, they had organized all they had seen into a formal administrative structure reaching from the *Audiencia* of Nueva Galicia on the north to the *Audiencia* of Lima on the south—from the Rio Grande del Norte to the Rio de la Plata. It was this routine application of a comprehensive administrative apparatus contiguously over the entire extent of their known new world which more than anything else geographically distinguishes the Spanish Empire. It was of course, as commonly interpreted, basically a case of Castile extended, the *Reconquista* enacted on an immense new stage, a continuity perfectly symbolized in the fact that Columbus first embarked in the very year that the last Moors were expelled from Iberia, so that three centuries of world empire followed hard upon six centuries of peninsular expansion.

The geographical extent of that empire was an exhibit of the remarkable range and power of individual energies, yet such men were the conscious bearers of a social system and they routinely imposed its structure upon every area entered. For the Spanish were much more than plunderers and the pattern of their empire was more a reflection of their desires to expand the realm of Christianity and to take root as a permanent social élite in command of land and labour. Thus, typically, a nucleus of Spanish as rulers and settlers was implanted within the bounds of a formal new town laid out central to a native population. Beyond the town a portion of the lands and certain rights to impress native labour were allocated. Meanwhile, missionaries worked systematically to convert *en masse* entire populations, which were to be integrated and eventually assimilated. As the range of such operations was extended outposts were established, some of which would themselves in time develop into more important centres. Thus, the whole process was a moving frontier of new communities, each of which was a cultural centre radiating Spanish civilization into the surrounding countryside.

Such a process operated within the larger framework of political territories, a hierarchy of districts, provinces, and vice-royalties, the whole matched in general though not always duplicated in detail by a hierarchy of ecclesiastical territories, from mission outposts to archiepiscopal sees.

This planned programme for the complete transplanting of a legalistic rigidly-structured Castilian society together with the programmatic transformation of native societies resulted, therefore, not in a simple binary pattern of conqueror and conquered, but a single, hierarchical, multi-racial, civic-centred landed society, made ever more cohesive by continuous acculturation and miscegenation. It was the application to the American scene of some basic patterns of medieval Iberia whose roots, in turn, reached directly back to the Roman colonization of Hispania.

It was an imperial system with a special geographical form, comprehensive and contiguous, a solid bloc of empire from the borders of Oregon to Patagonia (with the Philippines as a distantly detached but generally typical realm). The whole was subdivided into a regular territorial hierarchy, with each unit focused upon a Spanish-founded urban nucleus which served as the political and ecclesiastical centre. Such centres were imperial creations designed to foster and to serve a single integrated carefully structured society. All of these nuclei were connected by a simple network of prescribed traffic-ways, by which the whole of the Americas was brought into focus upon three ports (Vera Cruz, Cartagena, and Portobello) from which the strands of legal commerce were united in a single trunk to Spain (Figure 26).

The distinctiveness of such an imperial system becomes more

apparent when it is compared with that created by its Iberian neighbour in the other half of the globe during the very same years.

## THE SEA EMPIRES

The Portuguese capture of Ceuta in 1415 was the first sustained European foothold on the African shore and the first overseas base in what would become an imperial network reaching to the other side of the world. More locally and immediately, it opened the Strait of Gibraltar to European shipping and made Portugal pivotal to Italian and northern European influences, which in turn was probably decisive in making the Portuguese Empire in Africa and Asia a very different kind of empire than that of the Spanish in the Americas. For although eventually the Portuguese would seize most of the Atlantic ports of Morocco, they did not expand contiguously down the African coast but rather leapt from one useful base to another in a far-ranging search for profitable trade rather than for control of land and labour.

For a time, it was in general an orderly sequential expansion from one fruitful commercial region to another: from the Cape Verde Islands and Senegambia to the Guinea Coast, to Biafra and to the Congo. But their momentum soon carried them around the Cape into Indian waters and to the coast of India itself, so that while some of these African areas were still being probed, other Portuguese were ranging as far east as the China seas.

The rationalization of the Portuguese Indian Ocean empire under Albuquerque with a headquarters at Goa, trading stations along commercially profitable coasts, and strategic control of critical ocean passageways (as at Ormuz, Socotra and, later and briefly, Aden and Malacca) revealed what a very different sort of imperial system it was from that of the Spanish. For despite the presence of very active Christian missionary orders and the eventual emigration of some Portuguese colonists to a few locations, it was really only a network of scattered littoral points regionally focused upon some central insular base and all bound together by sea into a great trunk route, supported by oceanic island bases, to Lisbon. In such matters geographic form follows function, and the typical Portuguese base, on a small island or peninsula adjacent to an established entrepôt or passageway, reflected the very dominant interest in immediate commercial profit rather than in the political control of a native population and its land. To some extent, of course, it was a contrast which reflected different possibilities as well as motivations. In Asia and in parts of Africa, the Portuguese were in contact with well-developed societies which were not readily reduced by either arms or disease, whereas in Brazil the decimation of the native peoples soon led to the importation of African slaves and the development of wealth production directly under Portuguese control, developments which ultimately led to a colonial pattern not unlike that of Spain. Nevertheless, in general the Portuguese empire of the fifteenth and sixteenth centuries reflected in process and pattern more the precedent of Genoa and Venice than of the Iberian *Reconquista.*

The dominance of these early Italian commercial concepts of empire became much more pronounced in the seventeenth century when this Portuguese system was challenged in every major realm by other Europeans in a competition reminiscent of that between the Italian cities for the maritime East of their day. The resulting instability and complexity of pattern produced a set of imperial realms which well reflected the full geographical morphology attendant upon this particular system of empire.

The Dutch were the first of these newcomers to establish an extensive and more or less cohesive imperial system. In some instances they simply seized important Portuguese bases, as at Malacca and Amboina, but more generally they developed a duplicate system, similar in type and competitive in every Portuguese realm, from Brazil to the Moluccas (excepting only the western coast of the Indian Ocean where the Portuguese were being ousted by Omani maritime power), and intruding as well upon the Spanish in the West Indies. Similarly, the British and French soon established maritime systems of the same type in these same areas. Much less extensive but no less characteristic of the era and type were the trading stations of the Danes, the Swedes, and the Brandenburgers. All of these national systems were in direct competition for the same kinds of goods and, at times, in direct conflict with one another over particular locations.

Viewed globally, the competitions of these sea empires of the seventeenth and eighteenth centuries can be seen to be concentrated in four broad realms of the tropics each of which had its characteristic products, a pattern which may be summarized as shown in Table 1.

Not only were these empires primarily the products of commercial quests, they were very largely the products of companies rather than governments directly, a feature which gave an unusual degree of flexibility to their patterns. For a commercial company

TABLE 1

*Sea Empires, Seventeenth and Eighteenth Centuries*

| *Realms* | *Chief Products* | *European Interests Involved* |
|---|---|---|
| 'East Indies'—Malaysian peninsula and archipelago | Spices, medicinals, dye-stuffs, woods, sugar, 'China goods' | Portuguese, Dutch, French, English |
| India—especially Cambay, Malabar, and Coromandel coasts, and Bengal | Textiles, metalwork, lapidary, spices, dyestuffs | Portuguese, Dutch, French, English, Danes |
| 'Guinea'—west coast of Africa from Cape Verde to Cape Lopez | Slaves, gold, ivory, feathers | Portuguese, Dutch, French, English, Danes, Swedes, Brandenburgers, Spanish |
| 'West Indies', Guiana, and Brazil | Sugar, tobacco, cotton, dyestuffs | Portuguese, Spanish, Dutch, French, English, Danes, Swedes |

only wealth-producing positions were of basic interest, whether these be merely depots for the bartering of goods in foreign cities, a mineral district, or an agricultural plantation area. If the flow of wealth declined the position might be sold, traded, or abandoned. Although refreshment stations along the ocean routes, strategic naval bases, and political control over local areas were often considered necessary to the viability of the system, such positions tended to be considered part of the 'overhead costs' of commerce and not of intrinsic value. Thus rarely was colonization by European emigrants encouraged, and when it was there were usually attempts to control rather closely the numbers, area, and activities, as, for example, in the sustained but ultimately unsuccessful efforts of the Dutch company to restrict its original tiny colony of 'boers' to the immediate vicinity of Cape Town. In general, the Europeans of these imperial systems were no more than temporary residents in foreign lands, sojourners not settlers.

The small island lying just off a productive mainland was the ideal geographic base for such a system: secured by its natural moat and easily defended by a small garrison, readily accessible alike to rivercraft, coastal, and ocean vessels, and its obvious physical separateness making it easier to detach as a separate political unit or at least to maintain some greater degree of separation between European and native peoples and thus to minimize petty conflicts by minimizing local contacts. Bissau, Zanzibar, Bombay, Penang, Singapore, and Hong Kong come quickly to mind as obvious examples. So, too, the small fertile island was the ideal geographic base for tropical wealth production, for some of the same reasons. Where the labour was provided by slaves or an indentured class, however, it was best not to have a mainland too near (nor a backcountry on the island too large) in order to reduce the possibility of escape. Here the West Indies provided numerous good examples.

The local legal context of commercial trading bases varied in kind from no more than a rented building on the waterfront of an existing town, to a leased compound of residences, offices, and warehouses existing as a social enclave within or peripheral to a native city, to a wholly new community built up by Europeans on ground leased or granted by some local ruler, to a city, old or new or both, and some adjacent area, large or small, either purchased or seized by force, and completely under the jurisdiction of the European company. Indeed, any of these imperial systems, and often the pattern in any one region, very likely combined some or all of these types (and of lesser variations from them).

However, although these commercial companies usually operated under a board of governors which was principally guided by the interests of stockholders, they were also, in varying degree, under the general regulation of a king or parliament. Thus the geographical patterns of empire were also affected by wider national interests which might not always coincide with those of local commerce, and the political control of particular outposts might be changed abruptly by the outcome of military or diplomatic actions arising in some distant theatre. Thus, as an extreme example, the tiny island of Goree, lying in the shelter of Cape Verde strategic to the slave ship traffic, after having passed from Portuguese to Dutch and then to French hands by 1677, oscillated eight times between French and English control over the next century and a half, reflecting not so much any direct struggle over that particular locale as its role as one of many negotiable pieces in the global contentions of these two powers.

Thus, in summary, each imperial system was a series of local networks of interconnected coastal points bound together into oceanic trunk lines to the home base in Europe, with various holdings providing military and logistic support (Figure 27). In

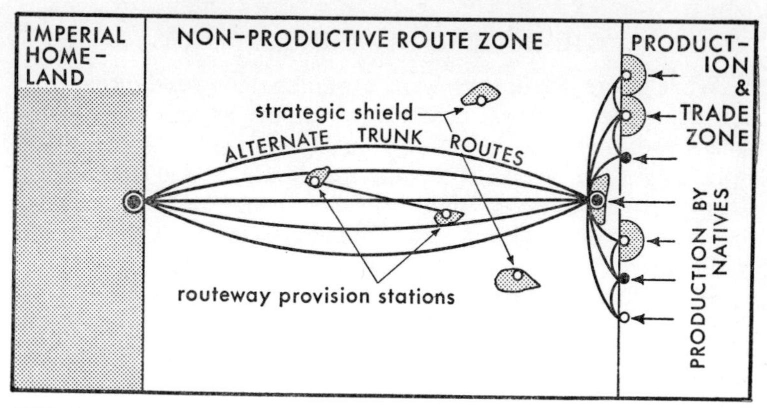


FIG. 27 Hypothetical sea empire

geographical form it was fragmentary, shallow in continental penetration, irregular in its territorial political hierarchy, and unstable in territorial pattern over any considerable period of time. The whole system fostered spatial and class segregation rather than integration of Europeans and local peoples; where much labour was needed, it often imported a subjugated labour force which in some places became the largest population or at least permanently complicated the racial and cultural character of those districts. Competition among the several national systems during these centuries resulted in a complex and unstable fragmentation of every broad tropical region of imperial interest (Figure 28).

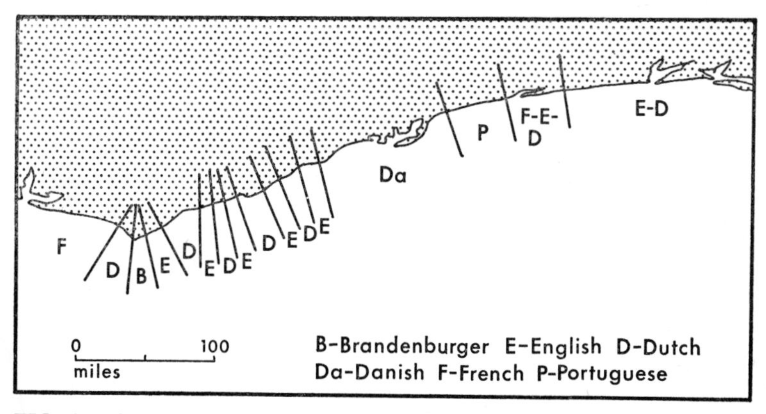


FIG. 28 Segmentation by sea empire rivalries, Guinea Coast, *c.* 1700

## THE BOREAL RIVERINE EMPIRES

While some Europeans were exploring the world ocean and spreading their imperial networks across the tropics, other Europeans were girdling the world in its northerly latitudes and spreading their networks across the broad boreal forestlands. Despite some obvious, and significant, contrasts between these continental and maritime networks there were also some striking similarities.

While the Portuguese were probing along the west coast of Africa, the government of Muscovy was sending expeditions across the northern Urals to reassert a Russian influence which had tenuously existed in Yugra since the earlier expeditions from Novgorod. The seizure of Kazan, a major mercantile focus of Eurasian trade, in 1552 opened a much broader front and set in motion a powerful eastward thrust of Cossack adventurers along the great boreal rivers which quite equalled in audacity if not quite in velocity that of the Portuguese over the tropical seas. Under the conditions of the time it was an astonishingly rapid expansion: the famous Yermak first led his motley band up the Kama and across the Urals in 1579; thirty years later similar bands were roaming the Yenisey valley; another ten years found them on the Lena in Yakutia; by 1638 they were on the shores of the Pacific, and ten years later had rounded the easternmost tip of the continent. In seventy years they had probed all the main valleys of northern Asia. In the century following, this great drive was extended beyond Asia and along the north-western coast of America, culminating geographically in Fort Ross and Bodega Bay, provision stations of the Russian-American Fur Company a few miles north of the Golden Gate.

This Russian thrust was matched in type and time by others westward across the full breadth of North America. Initiated by the French in Montreal and the Dutch in Albany, the American pattern was for some time retarded by political rivalries. The British succeeded the Dutch, then the French, then were themselves confined to the boreal lands with the loss of the United States. Only thereafter did they gain sufficient momentum to span the continent. Alexander Mackenzie reached the Pacific in 1793 and by 1811 the Northwest Company had a system of posts reaching from the St Lawrence to the mouth of the Columbia. It was a system briefly and weakly challenged on the Pacific slope by the

Americans, more critically by a rival British network extending westward from Hudson's Bay, which latter challenge led to an amalgamation which brought the whole of British America under the Hudson's Bay Company and brought the British and the Russians into immediate contact in the fiordlands of the American north-west.

All of these movements were powered primarily by the search for profits, chiefly in the form of furs obtained as tribute or trade from the sparse and primitive native populations. In early phases the Europeans were in many cases freebooters, sponsored perhaps by mercantile families or houses but completely unregulated. Only gradually were more orderly systems established.

In their local life and look—stockaded clusters of log structures dotted through the wilderness—and in their larger form and function these Eurasian and Euramerican systems were virtually identical, networks of summer camps, winter outposts, and district entrepôts, together with protective forts, provision stations, and more elaborate regional and system-wide depots and headquarters, the whole bound together as far as possible along the continental and coastal waterways. Many parallels with the sea empires are obvious: a far-flung commercial system of connected points through which a few men extracted wealth from the indigenous population, with forts guarding the vital river junctions and portages like those guarding the promontories and straits of the great sea lanes, and the farming posts on the few fertile spots of boreal wilderness as insular and important as the oceanic way-stations (Figure 29). So, too, when the furs of one district neared exhaustion, posts might be abandoned and efforts shifted to fresher fields.

Yet in other ways the difference between riverine and maritime was fundamental. For a river is a closed and fixed spatial system offering little of the flexibility or duplication open to those who competed on the high seas. The direct duplication of fur posts by British and Americans, as at Spokane and Kamloops, could be but a very brief interlude, with the outcome dependent upon which would control the lower reaches of the trunk stream, in this case the Columbia, the one feasible avenue between the interior and the sea. Thus these riverine systems were very largely monopolistic rather than competitive.

Furthermore, they tended to be contiguous rather than fragmentary. For even though these systems were at first and often

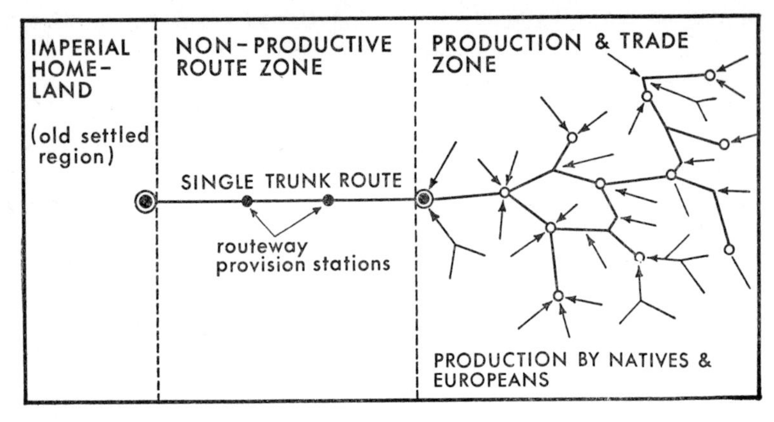


FIG. 29 Hypothetical riverine empire

long remained private and commercial, they were either accompanied by or themselves also served as agents of the national government. Thus all the lands their explorers and traders traversed were claimed and annexed to the body of the State whether valuable for furs or not, and their local populations were brought under some sort of administrative authority, whether immediately effective or not. In this relentless spread of a comprehensive unilateral stable administrative structure these riverine empires resembled that of Spain, yet here, too, there were important differences. For much of this vast boreal belt girdling the world in these latitudes was long unwanted by individual Europeans. Therefore most of those who came were sojourners in search of profits, not settlers in search of land and people which they could organize into a new society.

And yet, wherever a fur post was seated upon fertile ground, it did eventually become a nucleus for colonization. More importantly, these boreal regions were bordered all along the south by broad fertile regions which once colonized became thresholds for entry and ever-widening encroachment into these forestlands. Thus in each continent these broad realms became merely frontier portions of even larger political and economic realms, and lay open to gradual integration into the main body of the imperial State, an integration which transformed their native populations into captive minorities and which, indeed, transformed the whole from an imperial into a national system.

## SETTLER EMPIRES

This transformation of the boreal riverine empires can be viewed as the gradual encroachment upon and triumph of settler expansion over mere commercial expansion, a kind of imperial mutation, the one form developing gradually and almost inscrutably out of the other.

For the early fur posts in western Siberia along the southern margins of the boreal forest, such as Tyumen, Tobolsk, and Tomsk, were centres of a continuously expanding trans-Ural colonization by free peasants, a migration which would in time replace the native population and establish a solid wedge of Russian settlers between the Ugrian peoples of the forestlands to the north and the Turkic peoples of the steppelands to the south. By as early as 1700 the Russians outnumbered indigenous peoples over a broad belt of country. So, too, the earliest fur trade efforts in Canada were accompanied by the trans-Atlantic migration of permanent settlers to the lower St Lawrence Valley. Unlike the Russian, this French migration was rather sharply limited in numbers and time, yet it implanted a colony of such demographic vigour that it would spread a deeply-rooted peasantry densely and inexorably over a considerable area. Some of the other European enterprises on the American shore were initially similar in type or intention if not in scale: New Netherland and New Sweden, especially, both hoping to rely mainly upon trade but encouraging the emigration of a nucleus of colonists to bolster their hold and foster other forms of wealth production.

But there were other colonies of the same general source, area, and era which were much more completely representative of the settler type of colony, distinct in motivation and ultimately very different in form and result from more purely commercial expansions. New England, Pennsylvania, and Georgia are good examples, each a carefully planned attempt to transplant or establish some specialized form of European society, either a narrow selection of some existing part, such as Puritan or Quaker, or a careful design for some new ideal form, as in Georgia (or much later in the Wakefield colonies of New Zealand and South Australia). Rather different in purpose but eventually similar in general results were such commercial ventures as Virginia and the several penal colonies in Australia.

All of these were 'plantations' in the older sense, the colonization of new ground by permanent settlers, a process long apparent within Europe itself. The most obvious geographic difference from those precedents was their non-contiguous, trans-oceanic character, but there was also a major difference in relationships with local populations. For whereas, for example, the English colonization of Celtic lands, and the German colonization of Slavic lands was a slow movement in upon a peasant population which was already Christian and European in culture and thus to a very considerable degree resulted in an integration and often assimilation of the old population to the new, similar migrations into American and Australian lands were encroachments upon relatively primitive, semi-nomadic peoples, thinly and loosely spread upon the land, starkly different in culture, intractable to European control but highly susceptible to European diseases. The result therefore was very largely a replacement of the old by the new, through annihilation, drastic diminution, or expulsion.

It might be argued that where the natives were thus removed such movements were only momentarily truly imperial in the sense of the extension of political control over alien peoples. Nevertheless, it is appropriate to include these intercontinental settler migrations as a special type of imperial system. For not only were they always obviously imperial in their initial stages and for as long as a native non-European population remained a significant part of their local life, they also ineluctably evolved in some degree into a new kind of imperial relationship—that reflected in the political tensions between a mother country and its overseas (or, probably to a lesser extent, trans-montane) colonies. Such tensions inevitably arise not only from the obvious spatial separation, the difficulties of distance and the impossibility of maintaining close social communication between homeland and colony, but also from the differing levels of understanding of, and concern over, the inevitable new problems arising in the new colonial environment, and from the steady growth of a colonial-born population whose European ties are weakened with each successive generation. Such a compound of separations must eventually result either in an evolution toward greater autonomy or a revolution to abrupt independence (here again, the geographical continuity of the Russian settler empire allowed an exceptional result).

In all of these matters the United States, Canada, Australia, and New Zealand offer the most obvious examples. But there are numerous others which stand as variants of the type: Argentina and Uruguay, wherein very similar settler movements took place within an elemental framework which was a legacy of Spanish imperialism; Brazil in which a vigorous settler expansion, derived both from the coastal population of much earlier emigrants and from new migrants from Europe, spread broadly into the interior but everywhere mixing in some degree with the Indian population; South Africa, wherein Boer and later British settlers first expelled native societies from all the most fertile districts, then recruited Bantu labourers in such numbers as to leave the Europeans a minority population in many areas, especially in the mining districts.

The geographical character of a typical unit in this kind of imperial system would consist of a segment of coast upon which a European population had become firmly rooted and from which the native population had been eliminated; an inland frontier where the replacement of the one population by the other was still in process; a deeper zone, as yet beyond the reach of settlers but disrupted by an influx of natives displaced from the coastal area; and a remote interior unexplored but claimed in the provisions of a generous charter (Figure 30). The frontier—at once a line, a region, and a process, to use Turner's concepts—was the most distinctive geographical feature, dynamic and uncontrolled, continually changing the positions and proportions of all these zones. Once it had advanced into new environments and out of close communication with the older settled zone a geopolitical tension between frontier and seaboard mirrored in miniature that between overseas colony and mother country. The fact that in the best examples of an imperial system of such units, that of the British in eighteenth century America and in nineteenth century Australia, the units lay in an uninterrupted coastal series of segments, each a separate local political jurisdiction, with its own focus and internal network, yet contiguous (except for Tasmania and, in effect, Western Australia) and basically homogeneous in general cultural background, led to the gradual development of intercolonial networks of trade and communication, fostering a broader community of interests which might in time overbalance in importance the older trunk line connections across the seas to Europe. All of these features illustrate that once implanted, settler colonies took on a life of their own to a degree quite unparalleled by any other type of imperial holding.

Such an imperial system, comprehensive in claim and contiguous over broad areas, with European-founded towns as focal points, was in some ways similar in form to that of the Spanish, but it was quite different in its internal character, especially in its relations with native peoples and in its internal frontier—in that

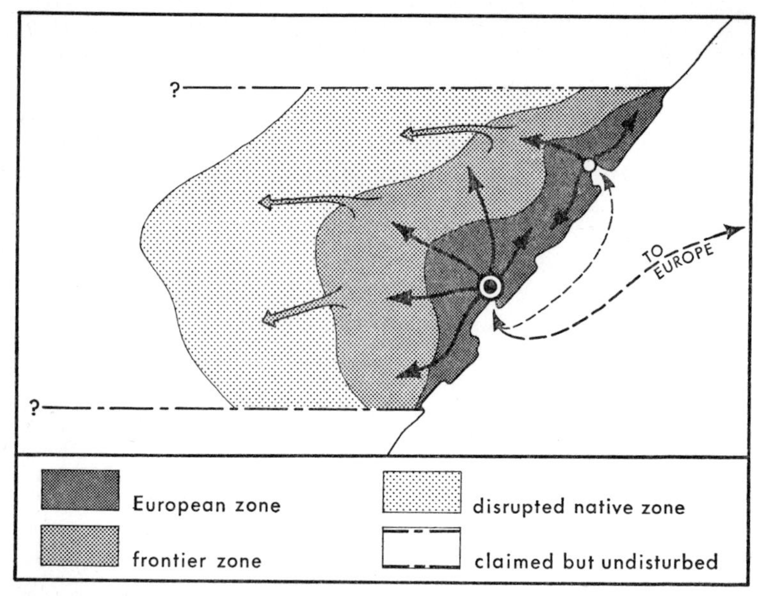


FIG. 30 Hypothetical colony of a settler empire

gradual, relentless, exclusive spread of Europeans upon the land. That permanent rooting of Europeans in conquered soil broadly along the seaboard made these settler empires obviously different from the commercial sea empires, and although the oceanic trunk line to Europe was important to both it was not absolutely vital to the settler colonies and the role of strategic holdings was a lesser part of such systems. Yet the political histories of Acadia, Quebec, and Cape Colony suggest that it was a difference not always appreciated by European diplomats, who shifted such well-established settler colonies from one flag to another as if they were no different from Goree, Mauritius, or Ceylon. The difficult, and at times ugly, complications of later years would make that difference starkly clear.

## NATIONALISTIC EMPIRES

A fifth and final type of empire was the last to appear, the most extensive in application, and the most purely 'imperial' in character. To label it a 'nationalistic' imperialism is to identify it with the outreach of virile national States, each highly conscious of power and position, strongly driven by the prestige of empire under the pressure of intense international rivalries. It was a drive also intimately related to the vigorously expansive stages of modern industrialism, with its omnivorous demand for materials, its need for expanding markets, and its development of ever more efficient tools for overcoming distance and for conquering peoples. It is therefore that form of imperialism freshest in mind, that of the past hundred years during which European States greedily gobbled up virtually the whole of the non-European world.

So extensive a development involving so many imperial powers in so many different culture areas was inevitably highly complex and no more than a brief note of a few salient features in each broad realm can be made, which together can give some basis for generalizations as to type.

In Tropical Africa we are dealing with the famous 'scramble' by which the entire realm was partitioned among a few powers. In part the resultant imperial pieces represented an extension inland of long-held coastal positions, dating back to the slave traffic of the sea empires, such as the creation of the broad blocs of Portuguese West (Angola) and East (Mozambique) Africa. Others were the work of commercial companies, now pushing inland in a new search for profitable trade and mineral districts (especially after the electrifying discovery at Kimberley), such as the Royal Niger Company and the British South Africa Company. Some of these operations were initially more the creations of individuals obsessed by some imperial vision than of the calculated strategies of national governments, as in the Rhodesias and, to some extent, the Congo. But most of this new colonial pattern expressed national rivalries for empire, quite regardless of any commercial value, as well exhibited by the Germans in Southwest Africa or the French conquest of the Sahara and the Chad. The overall result was the complete capture of Tropical Africa and its apportionment among half a dozen European powers in a complex pattern which bore virtually no relation whatever to indigenous culture areas.

In Southern Asia British India was certainly the dominant and most peculiar imperial creation: beginning as a series of coastal trading stations of the East India Company; expanding into a series of somewhat larger enclaves and spheres of influence following the defeat of rival sea empires for the Indian trade; then through power, alliances, and intrigue expanding into company rule over whole States; then, after control was shifted from the company to the crown, spreading relentlessly some form of British influence over the entire sub-continent; and culminating in the several campaigns into Afghanistan, not merely to quell local peoples on the Indian frontier but as part of a Eurasian-wide British strategy to block the rival expansions of Russia. The whole sequence reveals a gradual transformation from one type of imperial system to another with a corresponding alteration in geographical extent and form.

The Russian conquest of the pastoral and oasis societies of Central Asia was an excellent exhibition of nationalistic imperialism: the capture at great military cost and political strain of a vast realm full of alien recalcitrant peoples and of little discernible economic value. The subsequent formal division by Russia and Britain of Persia into spheres of influence separated by a neutral buffer zone, all without the participation of Persia itself, was another typical product of the intense rivalries of the era.

In Eastern Asia China represented still another kind of imperial realm. The first firm foreign encroachment, when the British in 1842 forced the cession of Hong Kong, the opening of five ports, and recognition of the concept of extra-territoriality, was an expression of the concepts of the commercial sea empires: a convenient but minimal base just offshore, a network of trading stations at profitable coastal points, and the maintenance of a sharp cultural segregation between peoples. Only near the end of the nineteenth century when imperial rivalries were greatly intensified was the new imperialism more fully expressed and even then, because the general territorial and governmental integrity of China was at least nominally supported, it followed to a considerable extent the earlier British example in the form of leased bases, special economic and political privileges, and spheres of influence. However, it became increasingly clear that the spheres were evolving toward little less than full imperial holdings and an outright dismemberment of China seemed imminent.

The fact that many of the islands of the Pacific were not formally claimed by European powers at the time of their discovery was a good expression of the dominance of commercial rather than national motivations of earlier imperialisms. The full nationalistic phase was initiated here as in Africa by the assertion of German claims to several groups, which quickly led to firm definition and considerable diplomatic conflict over various imperial holdings. The fact that the United States annexed only one island (Guam, as a waystation to the Philippines) of the Spanish Marianas was good evidence of that nation's very hesitant emergence as an imperial power; while the fact that Spain soon sold all the rest of her Pacific islands to Germany was evidence not only of a further stage in the decline of Spanish imperial power but also perhaps evidence that such a scattering of small islands was basically discordant with the Spanish concept of empire.

The broad realm of North Africa-Southwest Asia reveals still other peculiarities, most notably in Algeria. The French seizure of Algiers began as no more than an attempt to suppress chronic piracy and redress a diplomatic insult. But the French position evolved from the control of this one strategic base for the protection of the sea lanes to control of the entire coastline in order to secure other ports, to seizure of the entire fertile belt and the decision to transform Algeria into a settler colony—culturally and juridically into an extension of France itself—to military conquest of the High Atlas and the Western Sahara and the forcing of protectorates upon Tunisia and Morocco. What started as a minor and very practical punitive expedition ultimately resulted in a vast imperial realm reaching uninterrupted across the waste of Africa from the Mediterranean to the Congo. It was still another vivid exhibit of the mutation of imperial systems.

The impress of British control upon Egypt and her earlier acquisition of Aden, Perim Island, and Socotra may be classed as a strategy befitting a sea power, but her difficult conquest of Egyptian Sudan, the Italian seizure of Libya and Eritrea and abortive grasp for Ethiopia were archetypical of nationalist imperialisms, as was the last great scramble, under the guise of the Mandate system, for the Arab pieces of the crumbled Ottoman Empire in 1918.

The Italian conquest of Ethiopia in 1935 may be taken as the last event within the whole vast panorama of Western imperialism. It was undertaken as a recoup for a failure of forty years before and was so wholly and overtly a matter of national prestige, a thirst for the glories of a greater empire, that it produced a widespread revulsion of world opinion. But such a feeling came a little late to save the non-European peoples—there was really nothing left to grab.

The most obvious geographical feature of this era of nationalistic imperialism was its comprehensiveness. No inhabited island was too small or remote, no desert too broad or sterile to escape the claims of the imperial powers. Most of those few areas not directly subjugated were parcelled out as spheres of influence between competing imperial systems. The ubiquity and intensity of such rivalries was another striking feature, although the actual degree of partition varied greatly from one realm to another, with enormous consequences for the post-imperial era—consider, for example, the complications had India been parcelled about equally among the several powers which gained some foothold, or the differences had all of the Arab lands been amalgamated under a single imperial influence. Although every area was in some degree fragmented by these rivalries there was not the instability in national control so characteristic of the sea empires for most of the parcels were much larger and were held more out of national pride than for company profits. Thus during this era no holdings were abandoned, only a few remnants of earlier systems were sold (the Spanish Pacific Islands and the Danish West Indies), and one African island was traded for a more critical one in Europe (Zanzibar for Heligoland). For the rest, only defeat in a major war brought a shift in imperial control (Spain in 1898, Russia in 1905, Germany in 1918).

Figure 31 depicts a hypothetical morphology of an imperial holding and bordering units typical of this kind of imperial system. Capitals were most commonly on the coast, exhibiting in their position, functions, and character their role as the seat of power, focus of circulation, collector of wealth, and point of culture contact. Many were highly cosmopolitan, congregations not only of Europeans and local peoples but also of other non-Europeans—Chinese, Indians, Javanese, Lebanese, Armenians, recruited labourers or opportunistic migrants, important participants in European-generated economic developments. Only the richest and most accessible districts were exploited by Europeans, using local or imported non-European labourers. Beyond such areas the European presence was confined to a few political and military posts, agencies and garrisons strategic to native populations and troublesome frontiers. Frontiers had a special importance because every imperial holding was bordered by the holdings of other powers and because boundaries imposed by such rivalries very often severed traditional culture areas and thereby caused chronic problems of policing. The communications network reflected these strategic as well as commercial patterns. Backcountry areas of little economic interest were often no more than protectorates, ostensibly based on treaties of mutual support between the imperial power and a local leader. Further beyond might lie an area, usually one for some reason beyond easy practical entry, between expanding imperial systems and by agreement between

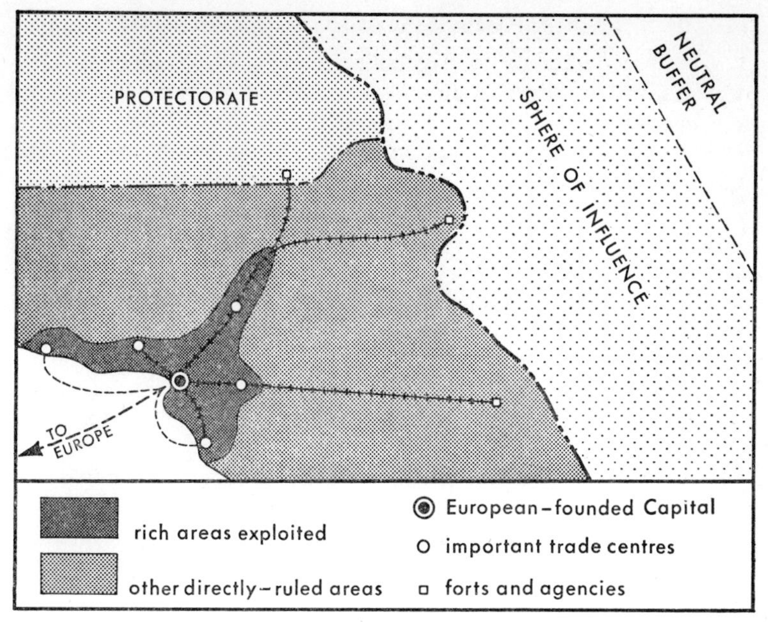


FIG. 31 Hypothetical patterns of a nationalistic empire

them divided into spheres of influence and a neutral zone in order to stabilize imperial rivalries.

The whole pattern exhibited that peculiar nationalistic obsession of controlling maximum territory rather than merely peoples or wealth-producing districts, and the intense rivalries gave a sudden acceleration to the whole process once it had begun in any region. The scale and variety of European-initiated developments might bring many thousands of Europeans to such a colony but they were almost entirely there as sojourners not settlers. Inevitably that same scale and variety involved large numbers of native peoples and thus an inevitable and often intense acculturation, yet the whole character of such an imperial system worked strongly to sustain a sharp social segregation of the two peoples. The very idea of such a nationalistic imperialism implied a strong feeling of inherent cultural superiority on the part of the imperial power. And of course it was that very display which provoked the deepest reactions and it was in these very cosmopolitan centres that the seeds of nationalism took deepest root amongst those very peoples most directly affected by the imperial presence.

## SOME GEOGRAPHICAL LEGACIES OF EMPIRE

Common reference to the 'collapse' or sudden 'dissolution' of imperialism or (more commonly) 'colonialism' is a reflection of the intensity and 'volume' of change during the last two decades. It is useful, however, to remember that these events are but the most recent phase of a general process which began with the American Revolution and continued sporadically thereafter. Thus although very nearly every part of the world has been at some time under some degree of rule by European peoples, European imperial States together never ruled the whole non-European world at any one time. Overall, expansions in some areas were accompanied by contractions in others. Any general view of the legacy of imperialism, therefore, must reach back to include those patterns produced a hundred and fifty or more years ago in the Americas.

Perhaps the most obvious geographical legacy is simply the distribution of sovereign States in the non-European world, for the great majority still exhibit the exact boundaries established by the former imperial powers. And within that overall pattern among the most striking features are those realms characterized by many small and even tiny States, such as Tropical America and the West Coast of Africa. Such areas of intense geopolitical complication are the direct result of intense rivalries dating back to the early years of the sea empires. It is a legacy of immense significance not only locally but to the world at large, as for example in the impact of so many small States upon the functions and character of the United Nations.

A more important matter for each State and its neighbours is the degree of geographic concordance between the most important culture areas and the political areas created by the imperial powers. The following simple set of logical categories provides an elementary framework (further illustrated by the hypothetical patterns of Figure 32), together with some specific examples:

1. *Cultural unity and imperial unity*

   There are few good examples: Cambodia as a French imperial province embracing most of the Khmer peoples; Morocco as long a rather separate Berber-Arab culture area, and perhaps a few of the southern African protectorates.

2. *Cultural unity and imperial diversity*

   The Arab lands, Malaysian core, Somaliland.

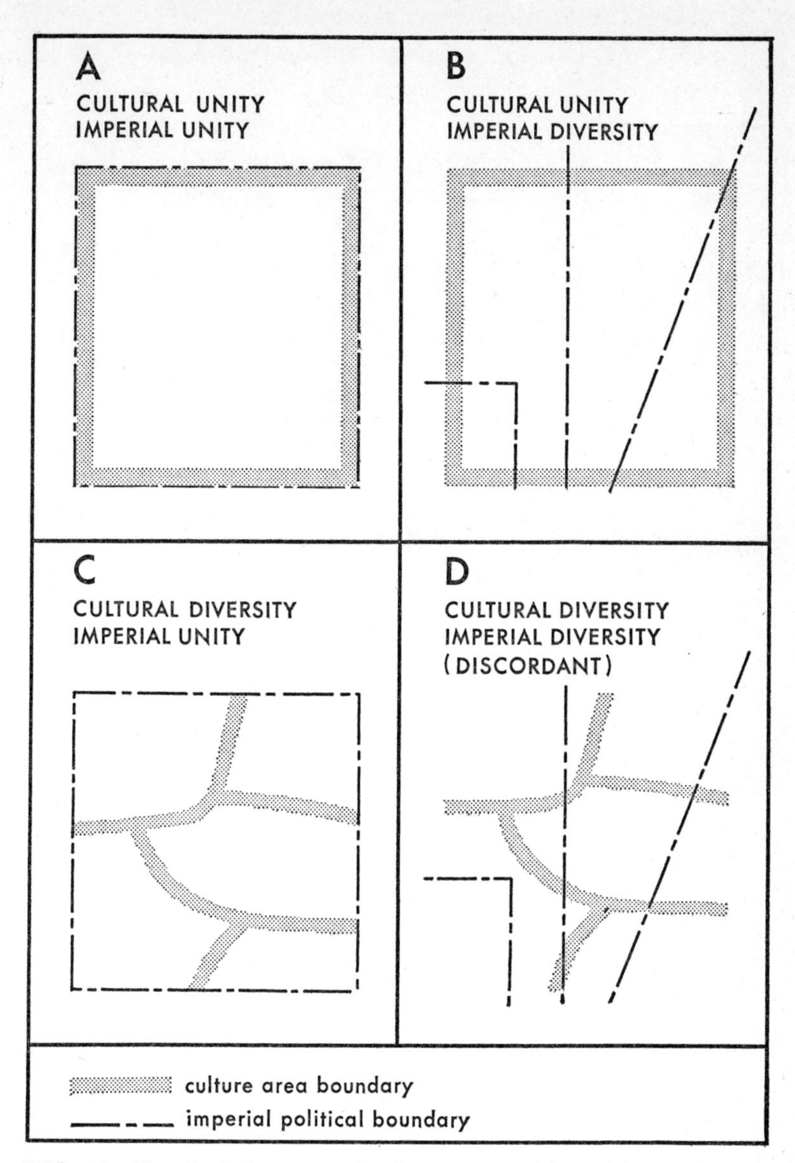


FIG. 32 Hypothetical patterns of culture areas and imperial areas

3. *Cultural diversity and imperial unity*
   e.g. British India, Burma, Nigeria, Canada, U.S.S.R.

4. *Cultural diversity and imperial diversity, but discordant in patterns*
   e.g. Tropical Africa, where tribal areas were complexly fragmented by rival imperialisms.

Such a scheme may be a useful tool for establishing some order out of a seeming chaos of geopolitical issues. For each specific case, the degree of previous cultural unity and the degree to which acculturation from the imperial power has altered such patterns must be examined. For example, the immediate amalgamation of the former British and Italian sections of Somalia suggests that the effect of differing imperial rules upon that pastoral society was relatively shallow. But such unification of diverse imperial parts is so far rare indeed. Despite a deep sense of cultural unity, the Arab States have as yet effected no firm amalgamations of their many political units. That divisiveness is certainly not alone a result of their having been parcelled among three different European empires, but that imperial heritage has certainly enhanced and further complicated older interregional rivalries. The imperial division of the Malaysian core area into British and Dutch sectors has also left a deep mark.

In the settler empires the very arrival of European colonists created a new cultural diversity. Where the native population was eventually reduced to relative insignificance, as in Australia and the United States, political and cultural patterns became essentially concordant. In New Zealand the persistence of the Maoris sustained a more significant diversity. Canada is the most notable exhibit of diverse European immigrant cultures being brought under a single imperial rule, while South Africa and the Soviet Union exhibit more complex diversities of Europeans and non-Europeans produced by a succession of imperial types but eventually encompassed within a single imperial rule.

The patterns of modern Spanish America, with its broad cultural similarity but fragmentation into eighteen sovereign States, are a clear exhibit of its peculiar imperial past. That cultural unity is of course a product of its particular form of imperialism, and that political diversity is a direct reflection of the territorial patterns of the administrative hierarchy of that empire. For in the 1820s the population was so clustered in a series of separate nuclei and communication between these was so difficult that upon the removal of centralized Spanish rule the imperial area broke into its administrative parts and most of the individual provinces, each with its capital and population cluster, became independent. The even greater fragmentation in Central America represents the successful assertion of sub-provincial administrative districts to independent life. Thus despite a concordance of broad cultural and imperial unities, other factors of geographical diversity caused a disintegration of a contiguous imperial whole.

The most spectacular example of an imperial unity broken by cultural diversity is of course British India. Although that imperial unity was greatly complicated by the mixture of native- and directly-ruled States, the subsequent fracturing did not in general follow such lines but rather the geographic patterns of the main religious cultures in which nationalism took its firmest roots. The powerful secessionist movements in such large imperial units as Nigeria and former Belgian Congo also raise doubts as to whether in these cases the pre-imperial cultural diversity can be contained within their unprecedented political bounds. Burma represents a more viable but still difficult legacy. The imperial design was established primarily to encompass the solid bloc of Burmese on the lowlands, but eventually so extended into the peripheral highlands as to bequeath to the new independent State an array of dissident minorities. Examples of further variations are readily apparent but enough has been offered to emphasize the richness of the theme for geopolitical study.

If we turn from these rather gross forms to a closer look at internal patterns we can also begin to see the significance of this variety of imperial types and morphologies to the patterns of culture contact and change. Grenfell Price has already described and compared some of the results in four of the settler empires in *White Settlers and Native Peoples*. Would that he had been able to include the results in Siberia also and thus given us a comparative view of the boreal lands, wherein the riverine empires were to a considerable extent succeeded by settler imperialism. In *The Western Invasions of the Pacific and its Continents* he has also called attention to the range of changes resulting from sojourner imperialisms in that half of the world. As he well understood, the patterns of spread and the power of the impact of exotic diseases, plants, animals, religions, language, technology, social, political and a host of other influences are to be understood not merely on the basis of the intrinsic characters of the cultures in contact, but also by a close look at the *geography* of that contact, by particular spatial patterns considered together with the scale of the setting and proportionate relationships of the peoples involved.

It is a commonplace in the study of imperialism to compare the systems of the several imperial powers as to their political character and results. It is not so common, but quite as important, to make a variety of other kinds of comparisons. These morphologies offer an elemental framework which can serve both the study of imperialism narrowly, as a set of spatial political systems, and broadly, as spatial patternings which have profoundly influenced the processes of change over much of the world. At the very least, it is hoped that this brief scan of a very broad picture may serve to suggest both what rich promise the burgeoning field of imperialism holds for the geographer and what an important perspective the field of geography holds for the student of imperialism.

# *Sir Grenfell Price*

Kt., C.M.G., M.A., D.Litt.

## BIBLIOGRAPHY 1918-1967


1. *Causal Geography of the World.* Adelaide, Rigby, 1918.
2. *South Australians and their Environment.* Adelaide, Rigby, 1921.
3. *The Foundation and Settlement of South Australia, 1829-1845.* Adelaide, Preece, 1924.
4. 'The Settlement of South Australia', *ANZAAS,* Vol. 17, 1924, pp. 439-48.
5. 'Geographical Problems of Early South Australia', *RGSA,* S.A. Branch, Vol. 25, 1925, pp. 57-80, followed by facsimile pages.
6. Introduction, *in* W. A. Cawthorne's *The Kangaroo Islanders.* Adelaide, Rigby, 1926.
7. 'South Australian Efforts to Control the Murray', *ANZAAS,* Vol. 18, 1926, pp. 444-56.
8. 'The Work of Captain Collet Barker in South Australia', *RGSA,* S.A. Branch, Vol. 26, 1926, pp. 52-66, with map.
9. 'The Historical Geography of the Northern Territory to 1871', *ANZAAS,* Vol. 19, 1928, pp. 282-93.
10. 'Sturt's Voyage down the Murray—the last stage', *RGSA,* S.A. Branch, Vol. 28, 1928, pp. 46-50, with maps.
11. L. Dudley Stamp and A. Grenfell Price. *World—A General Geography.* London, Longmans Green, 1928.
12. *Founders and Pioneers of South Australia,* Adelaide, F. W. Preece, 1929.
13. 'The Founders of South Australia', *RGSA,* S.A. Branch, Vol. 29, 1929, pp. 46-57.
14. [Address on] Captain James Cook, *RGSA,* S.A. Branch, Vol. 29, 1929, pp. 22-9.
15. The History and Problems of the Northern Territory, Australia (the John Murtagh Macrossan Lectures, Uni. of Queensland, 1930).
16. 'Extracts from a Journal of a Voyage in His Majesty's Ship "Buffalo" from England to South Australia', *RGSA,* S.A. Branch, Vol. 30, 1930, pp. 21-73.
17. *The Menace of Inflation.* Adelaide, F. W. Preece, 1931.
18. *The Progress of Communism.* Adelaide, F. W. Preece, 1931.

19. 'Pioneer Reactions to a Poor Tropical Environment: A Journey through Central and North Australia, 1932', *Geographical Rev.*, Vol. 23, 1933, pp. 353-71.

20. Experiments in Colonization, in *Cambridge History of the British Empire*, Vol. 7, Part I, pp. 207-42. Cambridge, Cambridge University Press, 1933.

21. *The Growth of Public Administration in the United States and its Lesson to the Australian Commonwealth.* Adelaide, Institute of Public Administration, 1934.

22. 'White Settlement in Saba Island, Dutch West Indies', *Geographical Rev.*, Vol. 24, 1934, pp. 42-60.

23. 'Early South Australian Maps in London', *RGSA*, S.A. Branch, Vol. 35, 1935, pp. 82-92.

24. 'The White Man in the Tropics', *Med. J. Aust.*, Vol. 1, 1935, pp. 106-10.

25. *The White Man in the Tropics.* Adelaide, Blennerhassett's Commercial Educational Society of Australasia, 1935.

26. 'White Settlement in the Panama Canal Zone', *Geographical Rev.*, Vol. 25, 1935, pp. 1-11.

27. 'The History of the College and College Register 1925-1935', *St Mark's College Record*, Special Supplement, Adelaide, Hassell, 1935.

28. A. Grenfell Price and Others, Joint Editors. *Centenary History of South Australia.* Adelaide, *RGSA*, S.A. Branch, 1936.

    *Contributions to Publications*:

    A. Grenfell Price and Mrs W. R. Birks.

    *Explorers by Sea and Land*, pp. 30-43.

    A. Grenfell Price.

    *Prospectus and Flotation 1829-1936*, pp. 44-56.

    *Pioneering Difficulties*, pp. 57-70.

    *Northern Territory*, pp. 95-101.

    *Bibliography* [*of S.A.*], pp. 396-401.

29. 'Geographical Problems in the Founding of South Australia', *RGSA*, S.A. Branch, Vol. 36, 1936, pp. 57-65.

30. [Address at the Pilgrimage to Light's Grave], *RGSA*, S.A. Branch, Vol. 37, 1937, pp. 75-6.

31. *Libraries in South Australia.* Report of an Inquiry Commissioned by the South Australian Government. Adelaide, 1937.

32. Foreword *to* 'Additional Secret Instructions Issued to Lieutenant James Cook 1768', *RGSA*, S.A. Branch, Vol. 38, 1938, pp. 99-100.

33. 'The Mystery of Leichhardt: the South Australian Government Expedition of 1938', *RGSA*, S.A. Branch, Vol. 39, 1939, pp. 9-48.
34. Presidential Address, *RGSA*, S.A. Branch, Vol. 39, 1939, pp. 1-8.
35. *White Settlers in the Tropics.* American Geographical Society, Special Publication No. 23, 1939. New York.
36. *What of our Aborigines?* Adelaide, Rigby, 1943.
37. 'The Comparative Management of Native Peoples—America, New Zealand, Australia', *J. Proc. R. Aust. Hist. Soc.*, Vol. 30, Part V, 1944, pp. 293-8.
38. *Australia Comes of Age.* Melbourne, Georgian House, 1945.
39. *The Collegiate School of St Peter, 1847-1947.* Adelaide Advertiser, 1947.
40. *White Settlers and Native Peoples.* Melbourne, Georgian House, 1949.
41. 'The Geopolitical Transformation of the Pacific and its Present Significance', *RGSA*, S.A. Branch, Vol. 52, 1951, pp. 1-12.
42. 'St Mark's College Scientific Work at Fromm's Landing', *RGSA*, S.A. Branch, Vol. 53, 1952, pp. 25-7.
43. The Social Challenge, in *Northern Australia—Task for a Nation*, pp. 179-209, Australian Institute of Political Science, Sydney, 1954.
44. *William Light.* Proc. Aust. Planning Cong., Vol. 3, 1954.
45. *Lake Eyre, South Australia: The Great Flooding of 1949-50.* Report of the Lake Eyre Committee. Adelaide, *RGSA*, S.A. Branch, 1955.

    *Contributions to Publication*:

    Introduction, pp. ix-xi.

    Historical Geography, pp. 1-5.
46. The Relations of White Settlers and Aboriginals in the Borderlands of the Pacific. Summary of Communications to the (Royal Anthropological) Institute. *MAN*, Vol. 55, 1955, p. 166.
47. *Report of the National Library Inquiry Committee, 1956-57.* Canberra, Commonwealth Government Printer, 1957.
48. 'Moving Frontiers and Changing Landscapes in the Pacific and its Continents', *Aust. J. Sci.*, Vol. 19, 1957, pp. 188-98.
49. Editor: *The Explorations of Captain James Cook in the Pacific as told by selections of his own Journals, 1768-1779.* New York, The Limited Editions Club, 1957. Another Edition: N.Y., The Heritage Press, 1958.
50. 'The Exploration of South Australia', in *Introducing South Australia*, pp. 32-6. Melbourne, Melbourne Uni. Press, 1958.

51. *Western Influences in the Pacific and its Continents.* Adelaide, Australian Humanities Research Council, 1959 (see Australian Humanities Research Council, Annual Report, No. 3, pp. 17-35).

52. Editor. 'The Humanities in Australia: A Survey with Special Reference to the Universities'. Sydney, Angus and Robertson for the Australian Humanities Research Council, 1959.

53. 'The Winning of Australian Antarctica', *RGSA*, S.A. Branch, Vol. 61, 1960, pp. 13-20.

54. *The Role of the National Library in Developing the Library Resources of the Nation.* Melbourne, 1961.

55. 'Captain James Cook's Discovery of the Antarctic Continent', *Geographical Rev.*, Vol. 51, 1961, pp. 575-7.

56. *The Winning of Australian Antarctica—Mawson's 'Banzare' Voyages.* Sydney, Angus and Robertson, 1963.

57. *The Western Invasions of the Pacific and its Continents: A Survey of Moving Frontiers and Changing Landscapes, 1513-1958.* Melbourne, O.U.P., 1963.

58. *The Importance of Disease in History.* Adelaide, Libraries Board of South Australia, 1964.

59. *The Challenge of New Guinea; Australian Aid to Papuan Progress.* Sydney, Angus and Robertson, 1965.

60. 'Further Notes on Cook's Possible Sighting of the Antarctic Continent', *Geographical Rev.*, Vol. 56, 1966, pp. 283-5.

61. *A History of St Mark's College, University of Adelaide and the Foundation of the Residential College Movement.* Adelaide. Council of St Mark's College, 1967.

# *The Contributors*

Bauer, F. H. is Professor of Geography-Anthropology at the California State College at Hayward, California. In 1952 he was a Fulbright Research Scholar to the Department of Geography at the University of Adelaide. It was from here that he did his field work for his Ph.D.

Cochrane, G. Ross is Senior Lecturer in Geography at the University of Auckland, New Zealand. While lecturing in the Department of Geography at the University of Adelaide in the 1950s he shared the teaching of a course on western invasions of the Pacific with Sir Grenfell.

Gale, Fay is a Lecturer in Geography at the University of Adelaide and was the first Ph.D. student to graduate from this department under the supervision of Sir Grenfell.

Hefford, R. K. is a Senior Lecturer in Economics at the University of Adelaide. He graduated from the first full degree course in geography given by Sir Grenfell and subsequently went to the Research School of Pacific Studies, A.N.U., to undertake research in economic development in South and South-east Asia.

Howlett, Diana is Associate Professor in Geography at the State University College, Oneonta, New York. She graduated originally from the Geography Department in Adelaide before going on to the Australian National University to pursue her interest in New Guinea and the Pacific which had been initiated by Sir Grenfell.

Lawton, Graham H. is the Professor of Geography at the University of Adelaide and became the first full-time head of this department in 1951.

Lea, David is a Senior Lecturer in Geography at the University of Papua and New Guinea. He was a student in residence at St Mark's College when Sir Grenfell was Master. He too was stimulated to follow his present interests by Sir Grenfell's course in Pacific Geography.

Maegraith, Brian is Dean of the School of Tropical Medicine and Professor of Tropical Medicine in Liverpool, U.K. He first graduated from the Adelaide Medical School. His association with Sir Grenfell began at St Peter's College. He was also a student in residence at St Mark's College when Sir Grenfell was Master.

Marshall, Ann is a Senior Lecturer in Geography at the University of Adelaide. She was co-lecturer with Sir Grenfell in establishing the Adelaide Department of Geography.

Meinig, D. W. is Professor and Chairman of Geography at Syracuse University, New York. In 1958 he was a Fulbright Research Scholar at the Adelaide Department of Geography and he offers his essay in tribute to the interest and encouragement which Sir Grenfell gave to his research on the South Australian wheat frontier.

Williams, Michael is a Senior Lecturer in Geography at the University of Adelaide. He is a historical geographer and has continued with the work on South Australian historical geography pioneered by Sir Grenfell.

# Index


Abelam people 173-184
Aborigines (Aust.) 65-88
  age and sex structure 78-80
  European contact 65, 66, 69, 71, 74, 80, 82
  full-bloods 65, 70, 71, 87
  genealogies 66, 67, 68
  mixed-bloods 65, 70-72, 82, 87
  reserves 65, 72-74, 77, 85, 87. *See also* fertility, mortality, population.
acculturation 218, 234, 235, 238, 239
agriculture, commercial 11, 14, 15, 18, 19, 21, 22, 25, 26, 28, 31-33, 35-38, 40-44, 120, 145, 192-195, 198, 199
agriculture, plantation 192-196, 221
agriculture, shifting 119, 174, 175, 187, 197
agriculture, subsistence 105, 118, 173, 174, 181, 195-200. *See also* cash crops, monoculture.
Apartheid (South Africa)
  economic effects 100-106, 110
  economic feasibility 106-109
    and employment 98, 100-103, 108, 109
  history of 89, 93, 94
    and politics 94-96, 107, 110, 111
  social effects of 94, 97-100. *See also* assimilation, integration, segregation.
*ara*. *See* kinship.
aridity. *See* climate—aridity, drought.
Asiatics 89-92, 95-109, 113, 115, 116, 121-124, 139, 142, 145, 162, 186, 190, 209
assimilation 69, 71, 72, 74-77, 87, 93, 217, 228

bananas 118, 175-177, 198
Bantu 89-92, 95-109, 113, 229
bride-price 181
Burke, R. O'H. and Wills, W. J. 55
bush fallow system. *See* agriculture, shifting.
bushland vegetation 117, 122-124, 128-131, 135-143

carriers (of diseases) 150-154, 156, 157
cash crops 119, 196, 198-200
cattle 40, 55, 58, 62, 97, 122-124, 139
census 65, 82
cholera 149, 156, 203
Christianity 214, 216-219, 228
climate 51-63
  aridity 52, 56, 61
  flooding 8, 9, 55, 56, 58, 60, 62, 127
  rainfall 29, 35, 51, 52, 54-62, 195
  seasonality 52, 54-56, 59, 61, 63, 117, 120, 126
clover (subterranean) 37, 45
colonialism, European 185, 188, 192-195, 208, 214
colonies. *See* specific names, e.g. Fiji, Papua-New Guinea. *See also* colonization, European, imperialism, Western.
colonization, European 1-4, 153, 185-191, 208-209, 216, 219, 221, 226-228, 230, 238. *See also* colonialism, European, settlement, European.
Coloureds 89-92, 95-109, 113
commerce 101, 102, 206, 216, 218-222, 224-227, 230-232, 234. *See also* trade.
commercial agriculture. *See* agriculture, commercial.
communism in Africa 96, 98
conservation (soil, vegetation) 137-139, 143, 145
copper. *See* mining.
crime 98, 99
culture areas 231, 234, 236, 237, 239
culture conflicts 93-97, 121, 139, 222, 228, 232
culture contact 65-69, 86, 87, 94, 115, 120, 187, 191, 217, 228

*dalo*. *See* taro.
deficiency diseases 173
demography. *See* population.
desert and semi-desert 10, 52
diet 160, 161, 173, 174, 181, 203
discovery and exploration 4, 54, 55, 61, 214-218, 223, 224, 226
diseases
  control of 106, 203, 228
  endemic 151, 152, 155, 156, 169
  in animals 150, 161, 162, 164

diseases—*continued*
  spread of 70, 72, 84, 86, 87, 89, 105, 149, 151-160, 162, 164, 165, 168-170, 188, 220, 239. *See also* particular diseases, e.g. malaria.
drainage (artificial) 28, 29, 33, 35, 46. *See also* swamps.
drainage (natural) 9, 17, 126-136, 138, 141, 142
drought 32-34, 36-38, 42, 43, 62. *See also* climate—aridity.
dry farming. *See* farming, dry.
dual economy 107, 108, 192, 203

education 99, 100, 189, 202, 203
emergent nations 150, 152, 154, 155, 157, 163, 166, 169
empires
  continental 216-218
  nationalistic 230-235
  riverine 223-227, 239
  sea 215, 218-222, 230-232, 234, 236
  settler 227-230, 238
employment 59, 98, 100-105, 108, 109, 195, 200, 201, 206-208. *See also* labour.
epidemics 72, 151, 153, 154, 162. *See also* disease and particular diseases, e.g. malaria.
erosion. *See* soil erosion. *See also* land slips, sheetwash.
European colonization. *See* colonization, European. *See also* colonialism, European.
European settlement. *See* settlement, European.
exchange of food. *See* food—exchanges.

families, Aboriginal 73-76. *See also* kinship.
farming, dry 37-41, 43, 47
feasts. *See* food—exchanges.
fertility 69-77, 80-83, 87, 88
  birth-cohorts 75, 76, 80-86
  birth-rates 69, 71-73, 80, 87, 88
Fiji 115-147
Fijians 115-118, 120, 121, 123, 124, 130, 145
firing (vegetation) 122-124, 126-130, 136, 138-143, 145
flooding 8, 9, 55, 56, 58, 60, 62, 127. *See also* climate—rainfall, drainage, swamps.
food 173-184
  distributions 177, 178
food—*continued*
  exchanges 174, 177-181, 182, 198
  payments 175, 176, 187
  plantings 177, 179, 181
  surpluses 174, 181, 182, 198
  taboos 173, 177
forest vegetation 7-9, 13, 15, 117-119, 120, 122, 124-126, 128, 131, 135, 138-145
forestry 117, 119, 120, 137-139, 141-145
frontiers. *See* moving frontiers.
fruits. *See* tropical fruits.
furs 225-227

gasau 119, 123, 129, 139
genetic selection 159, 160
geopolitics 213, 217, 218, 221, 222, 229, 232, 236, 238, 239
gold. *See* mining.
Goyder's line 28-30, 39-42, 45
granite 124, 126, 129, 130, 133-135
grasslands 6, 9, 19, 54, 56, 117, 122-126, 128-145
grazing 122, 124, 126, 128, 129, 137, 139. *See also* cattle, pastoral lands, sheep.
Gregory, A. C. 54
gross domestic product 110, 191
guava 126, 137, 142-144
Gulf country. *See* Queensland, North-West, pastoral lands.

health 92, 105, 106, 149-171, 172, 189, 203
heartlands 217
historical geography 2, 3, 47, 214
hosts (to diseases) 151, 153, 159, 162, 166
houses, tambaran. *See* tambaran houses (N.G.).
humidity 117
hundreds (land) 2, 14-16, 18-22, 26, 32, 41, 43, 45, 46
hunger 173

imperialism, Western 188, 192, 213-240. *See also* colonialism, European, colonization, European, empires.
income. *See* per capita income.
independence. *See* self-government.
Indians (in Fiji) 115, 116, 121-124, 139, 142, 145
industrialism 100-104, 108, 109, 150, 200, 201, 231
infant mortality. *See* mortality—infant.

infections. *See* carriers, diseases, epidemics, and particular diseases e.g. malaria.
initiation ceremonies 177, 178
integration 109, 217, 223, 226, 228
international rivalries 149, 153, 188-190, 220, 222-225, 230-236, 238
irrigation 36, 40, 46, 154, 155

jobs. *See* employment.

kinship 74, 118, 177-180, 187, 198
kwashiorkor 105, 160, 161

labour 98, 107-109, 121, 154, 165, 187, 193-195, 197, 200, 202, 209, 217, 218, 222, 223, 229, 234
land clearing 1, 15-17, 22-24, 30, 39, 42, 43, 119, 120, 198
Landsborough, William 55
land slips 126, 127, 129-131, 133, 134, 136, 138, 144
land tenure 115, 118, 121, 187, 196-198
Leichhardt, Ludwig 54

MacGregor, Sir William (Administrator) 196
Malan, Dr D. F. 94, 95
malaria 105, 106, 151, 153, 157, 159, 161, 168, 187, 203
mallee vegetation 9-10, 16, 19, 21-22, 27, 30, 38, 40-43
malnutrition 89, 160, 161, 169, 173
Maori 73, 77
marginal lands 29, 30, 32, 33-35, 43, 45
marriage 75, 76, 94, 98, 181
migration. *See* population—movements.
mining 13-19, 120, 186
minority groups 65, 90, 94, 106, 113, 226, 229, 239
miscegenation 75, 76, 218
mission grass 123-126, 132, 138-145
missions 66, 67, 69, 72, 75, 87, 195, 217, 219. *See also* Christianity, Point McLeay.
mixed-bloods. *See* Aborigines—mixed-bloods, Coloureds.
monoculture 37, 44, 115
mortality 68, 69, 71, 73, 82-86, 87, 89, 105, 106, 151, 169, 203, 228
  age at death 68, 69
  death-cohorts 83-86

mortality—*continued*
  death-rates 69, 71, 72, 73, 86, 92, 203
  infant mortality 71, 83, 84, 86, 87, 92, 105, 203. *See also* population.
moving frontiers 214, 217
mullenizing 30, 47

nationalism 93, 96, 97, 111-113, 204, 205, 230-235, 239. *See also* self-government.
National Party (South Africa) 94-96, 98
native peoples. *See* settlement, European—effects on native peoples.
nomads. *See* population movements.
nutrition 160, 161, 173, 174, 198, 203

over-population. *See* population—increase.

Papua and New Guinea
  Australian policy 185, 189, 190, 192, 194, 195, 200-209
  economic development 190, 191, 193, 194, 196-201
  history 185, 187-190
  politics and government 204-209
  primary industries 192, 193, 195-200
  resources 186
  secondary industries 200, 201
  social conditions 202-208
parallel development. *See* dual economy.
parasites 151, 153-159, 162
pastoral lands 13, 18-22, 26, 28, 29, 32, 35, 54-56, 58-62, 194
payment in kind. *See* food—distributions, payments.
per capita income 104, 105, 190, 191
pigs 42, 176-181
pine plantations 138, 141, 143, 144. *See also* forestry.
plantation agriculture. *See* agriculture, plantation.
plantings (crops). *See* food—plantings.
plural society 89, 107, 108, 115, 121, 186, 203, 206, 207
Point McLeay Reserve 65-88. *See also* Aborigines (Aust.).

political geography. *See* geopolitics.
political rights 95-97, 107, 113, 121, 204-207
politics and government 21-30, 32, 33, 38-40, 42, 61, 94-98, 107, 110-112, 153, 204-209
population
  age groups 68, 78-80, 106
  comparisons 71, 73, 77-79, 82, 83, 89-92
  density 115, 116, 186, 195
  increase 72, 90-92, 116, 169, 170, 204
  movements 77, 149-155, 157, 159, 160, 163-169, 195, 215, 216. 222, 223, 227-229, 234. *See also* Aborigines (Aust.), fertility, mortality.
ports 17, 18, 21, 31, 215, 218, 219
Price, Sir Grenfell xiii-xviii, 121, 214, 239.

*qato* 126, 139-141
Queensland, North-West 54, 55, 61

racial discrimination 97-99, 104, 106, 108, 121, 207. *See also* apartheid.
railways 3, 17, 18, 25, 27, 31-33, 35, 38, 40-43, 47, 53
rainfall. *See* climate—rainfall.
reserves 65, 72-74, 77, 85, 87, 105
Ridley's stripper 14, 15
ritual and ceremonial. *See* food—exchanges, taboos.

sago 175-177
sandalwood 119
savannah vegetation 6-13, 117, 119, 122-123, 142-145
scrub clearing. *See* land clearing.
scrubland vegetation 22-24, 120-124, 130-132, 136-145
sealers 66, 86
seasonality. *See* climate—seasonality.
segregation 109, 223, 232, 235. *See also* apartheid.
self-government 94, 163, 185, 205, 228, 239
separate development. *See* apartheid.
settlement, European 1-4, 14, 51, 54, 89, 94, 119, 120, 121
  effects on native peoples 65, 66, 70-72, 84-87, 89, 93-97, 118, 119, 121, 152, 153, 164, 165, 216-220, 222, 225-229, 231-235, 238
settlement, European—*continued*
  effects on vegetation 1, 15, 17, 119-121. *See also* colonization, European.
settler colonies 120, 121, 214, 221, 226-230, 233, 235, 239
sexual behaviour 75, 76
sheep 8, 15, 45, 55, 58, 61, 62
sheetwash 126, 127, 129-131, 133, 134, 136-138, 144. *See also* soil erosion.
shell rings (N.G.) 179, 181, 188
shifting agriculture. *See* agriculture, shifting.
shistosomiasis 151, 154, 155, 162
slavery 188, 220, 222, 223, 231
sleeping sickness. *See* trypanosomiasis.
smallpox 151, 166, 168, 203
social welfare 97-106, 160, 161, 202-208
soil erosion 43, 123-127, 129-134, 136, 137, 139-144. *See also* land slips, sheetwash.
soil fertility 9, 24, 117, 195-198. *See also* trace elements.
soils 6-11, 13, 24, 32, 117, 123, 126-130, 133-135, 139-142
sojourner colonies 93, 121, 214, 221, 226, 227, 235, 239
South Africa
  economic sanctions against 110, 111
  immigration 92
  minorities 93
  political future 110-112
  settlement 93
  social conditions 97-106. *See also* Asiatics, Bantu, Coloureds, whites.
South Australia, settlement
  government policy 21-30, 32, 33, 38-40, 42
  phases of s.
    1836-1849, 11-14; 1850-1859, 14-19; 1860-1868, 19-25; 1869-1879, 26-32; 1880-1892, 32-37; 1893-1920, 37-43; 1920 onwards, 43-46
staple foods 118, 196-198
stock. *See* cattle, pastoral lands, pigs, sheep.
stump-jump plough 30, 47
subsistence agriculture. *See* agriculture, subsistence.
sugar cane 115, 117, 120-122, 124, 139, 142, 145, 198

superphosphates 10, 36, 38, 39, 41, 47
swamps 11, 17, 28, 32, 33, 35, 46

taboos. *See* food—taboos.
tambaran houses (N.G.) 176-179
Taplin, Rev. George 67
taro 118, 119, 175-177
tenant farmers 122
timber-getting 117, 119, 120. *See also* forestry, sandalwood.
Tindale, Norman B. 67, 68
township sites 2, 19, 21, 24-26, 30, 34, 38, 40, 46
trace elements 10, 45, 46
trade 13, 14, 18, 21, 25, 31, 106, 119, 120, 188
trade routes 215-219, 221, 229
traders 119, 120, 216
trading posts 188, 215, 217-222, 224-227, 231, 232
transport. *See* railways.
tropical fruits 118, 119, 175, 176, 187, 198
trypanosomiasis 151, 153, 154

under-development 163, 165, 185, 188-192, 195-197, 204, 208
under-nutrition 160, 174
United Nations Organization 110, 112, 186, 190, 192, 205
urbanization 74-77, 91, 98, 117, 160

vectors. *See* carriers (of diseases).
vegetation change 3, 8, 15-17, 119, 120, 122-124, 126, 128-130, 134, 136-145
vegetation regeneration 119, 120, 136-140, 144
vegetation types 6, 7, 8, 117-120, 122-132, 134-145
villages 118, 119, 121, 175-181, 187, 195-197
virus diseases 161, 164, 166
Vorster, B. J. 111

*wabi*. *See* food—exchanges.
Wakefield, Edward Gibbon 3, 4, 14, 24, 25, 227
Western invasions. *See colonization*. European, imperialism, Western.
whalers, 66, 86
wheat 11, 14, 15, 18, 19, 21, 22, 25, 26, 28, 31, 32, 33, 35-38, 40-44
whites 71, 77, 89-92, 95-109, 113. *See also* colonization, European, white settlers in tropics.
white settlers and native peoples. *See* settlement, European—effects on native peoples.
white settlers in the tropics 52, 61-63, 94, 121, 149, 150, 164, 189
wool 14, 18, 21
workforce. *See* labour.
World Health Organization 155-157

yams 118, 175-181